Advance Praise for *Fight Like a Mother*

"Too many mothers have been tragically impacted by gun violence, and yet they courageously turn their grief and frustration into action. I'm honored to fight alongside them, working in the Congress to pass effective, commonsense gun violence prevention reforms, and I'm so glad Shannon Watts has written this book and provided a road map for other everyday activists to make a difference."

—*Speaker Nancy Pelosi*

"There's never been more energy behind America's gun safety movement than there is today, and Shannon Watts and Moms Demand Action are a big reason why. Shannon is fearless, and her story shows that change—even on the toughest issues—really is possible."

—*Michael Bloomberg, entrepreneur, philanthropist, and three-term mayor of New York City*

"There is nothing more powerful than a mother on a mission—and Shannon perfectly explains why. This mother has fought to lay the groundwork for one of America's largest movements for gun violence prevention."

—*Congresswoman Lucy McBath (GA-06)*

"A page-turner, overflowing with moving stories as well as practical advice for those who are waking up to the realization that there is too much at stake to sit on the sidelines. . . . Shannon's courage and down-to-earth approach radiate from every page, and the theme of finding her 'soul sisters' in her fight is as powerful as it is universal."

—*Cecile Richards, former president of Planned Parenthood and* New York Times *bestselling author of* Make Trouble

"Shannon Watts is a true activist for our time. After Sandy Hook, she channeled her despair and fury into action. *Fight Like a Mother* will inspire you to take an issue on and, ultimately, change the world."

—*Katie Couric, journalist,* New York Times *bestselling author, cancer advocate, podcast host, and documentary filmmaker*

"Shannon Watts is the embodiment of an age-old story that never ceases to inspire: anyone, regardless of who and where they are in life, can create a paradigm shift. Even if change seems impossible. *Fight Like a Mother* is practical and personal—not only equipping you with the necessary and nitty-gritty tools to start a movement, but also the fearless mentality to do it now."

—*Preet Bharara, former US attorney and host of the* Stay Tuned with Preet *podcast*

"Shannon Watts's fight is for her kids and all our kids. That fight animates her work with Moms Demand Action, her life, and her wonderful book. In *Fight Like a Mother,* Shannon shows how she stewards this rocket ship of hope for those of us distraught by the epidemic of gun violence in our country through clear, disciplined approaches to movement building, volunteering, branding, and deep gun violence data literacy. A great read!"
—*Chelsea Clinton,* New York Times *bestselling author and vice chair of the Clinton Foundation*

"It's encouraging to see someone tell the story that, sadly, has unfolded over and over again in this country—moms springing into action when their children are put at risk. This book has the potential to become a primer on how to activate the parent community and make change so we can see a better tomorrow."
—*Soledad O'Brien, journalist, documentarian, news anchor, and producer*

"Shannon Watts is proof that activism and community can actually change laws and lives. *Fight Like a Mother* is a brilliant call to action for anyone hoping to get off the sidelines and create meaningful and lasting impact."
—*Sophia Bush, actress and activist*

"The world needs angry women right now. It's time we embrace the passion and action behind saying 'enough is enough.' *Fight Like a Mother* is a testament to the positive power of women's anger and using it as an unstoppable force for change."
—*Soraya Chemaly, writer, activist, and bestselling author of* Rage Becomes Her

"There is nothing more powerful than a mother's love, and there is nothing more terrifying than a mama bear protecting her cubs. Shannon and Moms Demand Action have revolutionized grassroots organizing; and they've successfully challenged and beat back one of the most powerful lobbying groups in DC: the NRA. This book is a guidebook to all who are looking to drive real and lasting change in their communities."
—*Congresswoman Robin L. Kelly (IL-02)*

"Shannon is an inspiration to every mom fighting to build a better, safer country for our kids. Her story reminds us that, together, we can be a powerful force for change—and that you should never underestimate a mom's ability to get things done."
—*Governor Gina M. Raimondo of Rhode Island*

"Have you ever thought you weren't powerful enough to make a difference? Well, Shannon Watts is proof that one fiercely dedicated person can impact an entire movement. *Fight Like a Mother* illuminates how the very skills that we use daily as mothers are the exact skills necessary to take down the NRA. Shannon is my personal hero. Read this book and she will be yours as well."
—*Debra Messing, actress, advocate, and activist*

FIGHT LIKE A MOTHER

FIGHT LIKE A MOTHER

How a Grassroots Movement
Took on the Gun Lobby and
Why Women Will Change the World

Shannon Watts
with Kate Hanley

HarperOne
An Imprint of HarperCollins*Publishers*

Shannon Watts will donate a majority of the books' proceeds to nonprofits working to end gun violence.

HarperCollins books may be purchased for educational, business, or sales promotional use. For information, please email the Special Markets Department at SPsales@harpercollins.com.

FIRST EDITION

Designed by Bonni Leon-Berman

Library of Congress Cataloging-in-Publication Data has been applied for.

ISBN 978-0-06-289256-0

19 20 21 22 23 LSC 10 9 8 7 6 5 4 3 2 1

CONTENTS

FOREWORD

I wonder how many memories start with a date. Certainly for me, the days my children were born, the day I was married, my birthday, my siblings' birthdays, my parents', the day my mother died—these dates are indelible.

But the ordinary days become indelible only in retrospect, in the recalling of the events of the day over and over, in the retelling of the story of what you felt and of how that day changed you.

On December 14, 2012, I was working in New York City. My job is erratic and takes me many places, and the effort it has taken to combine my family life with my working life (as it is for most parents) has been enormous. So that day in 2012, I was luxuriating in the privilege of a set job in Queens. I had reasonable hours, great pay, and a very short commute from my home in Manhattan. I was home for dinner every night. It was a dream job.

My kids were in the same school, but in different divisions—my fifteen-year-old son was in the upper school, and my ten-year-old daughter was in middle school. Their vacations were slightly different, so my daughter, Liv, had just started her holiday break, while my son, Cal, was still in class. My husband, Bart, was working that day as well. I had a light day on the set and wasn't scheduled to work until the afternoon, so I thought I would bring Liv to work with me and she could watch the few scenes I was shooting and then we would return home together. Once again, I thought how lucky I was that I had a job flexible and generous enough to allow me to bring my child to work.

Early that day, the news broke about the Sandy Hook Elementary School shooting. I felt shock, and disbelief, and horror. How could this happen? How could tiny children be shot in the safety of their own school? What could be done for those families, how will they go on? And then—how will I explain this to my own children?

Liv was going to be with me all day, and so I felt I could control the narrative and decided we would discuss this as a family, when we were all together at the end of the day. Her father and I could tell our children about the tragedy, and assure them that they were safe. But until then, I would keep the news away from her.

When the van came to get us for work, I whispered to the driver to please keep the radio off. In hair and makeup, I asked them to please turn off the TV, and I asked the other actors and crew members to please not mention anything in front of my daughter. Everyone was very shaken up by the news, and we all had difficulty concentrating that day, but everyone wanted to protect the little girl who happened to be visiting our set.

We returned home in the early evening and had time before dinner, so I put on holiday music, opened the dusty boxes of Christmas ornaments, and began decorating the tree. Liv helped for a little while but then became distracted by her newly acquired smartphone, which, I am loathe to admit, we had just given her. At that moment, she looked up from her very carefully monitored phone, with its selected numbers of grandparents, mom, dad, brother, and friends, and said, "Mommy, did a bunch of little kids get shot today?"

I was so ashamed. That was the moment I realized that I had failed as a mother. I wasn't keeping her safe. Attempting to shield a child from terrible news does nothing to prevent them, or any other child in the United States, from experiencing gun violence. Only by DOING something about gun violence was I going to be the responsible parent and citizen I wanted to be.

So I began speaking out. Traditionally, actors have been reluctant to talk about guns, because of the relationship in our culture between guns and entertainment. But one of the things that I learned was that the entire world consumes the exact same video games and movies, and we are the only developed country in the world that has this level of gun violence. I noticed that Michael Bloomberg had started an organization called Mayors Against Illegal Guns, and I sent money to it. I began speaking out in interviews. I read as much as I could about gun violence, and I started following people on Twitter who were speaking out about this epidemic. One of them was a woman named Shannon Watts.

Shannon Watts had a similar experience to mine on the day of the Sandy Hook shooting. Her sense of outrage led her to start a Facebook page initially titled One Million Moms for Gun Control. Within moments of her creation, incensed moms across the country joined her, and Shannon understood that she had tapped into a "tsunami of rage" among American moms. Just like that, Shannon Watts became a mother of the movement.

I didn't get to meet Shannon for another two years. By then, I had tired of the blowback I received when speaking out against gun violence and realized I needed to do more than send money and tweet. I had been avidly following Shannon and realized that her organization, now titled Moms Demand Action for Gun Sense in America, had partnered with Mike Bloomberg's group, now called Everytown for Gun Safety. Inspired by what they had done, I went to them and offered to create a nonpartisan council of people from my community (known for their very big mouths) who were willing to lend their loud voices to the gun safety movement. And so I founded the Creative Council at Everytown—our initial group consisted of everybody on my contact list who would answer my email. I was modeling my behavior on Shannon, who by this point had become my own personal hero.

When I did meet Shannon in person, she did not disappoint. She was exactly as she had been described to me—a mom of five, modest, a full-time volunteer who worked around the clock to change culture and legislation. She is an empath, an introvert, a quiet warrior, and a leader whose mission statement never fails to both make me cry and galvanize my activism—"If we have lost our children, we have nothing left to lose." In this wonderful book, Shannon explains what it is to be an activist and how all of us have the ability and the bandwidth to do more than we think is possible. She explains that moms have a unique toolkit of assets, and the very skill sets we use to manage our family lives are the ones that make us uniquely powerful activists.

Moms Demand Action has grown from a Facebook page to our nation's "first and largest grassroots counterweight to the gun lobby." I am proud to be among the now thousands of volunteers, along with my now teenage daughter, working to change the culture of gun violence in the United States. And I am very proud to know Shannon Watts.

—*Julianne Moore*

INTRODUCTION

On the morning of December 14, 2012, I was home alone—just me, the morning news, and a few baskets of laundry that needed folding. With my husband in meetings and my five kids in school—my son in middle school and the four girls in high school or college—I was looking forward to a little quiet time. I'd just dumped another pile of clothes onto the bed for folding when the news broke of a shooting at Sandy Hook Elementary School in Newtown, Connecticut.

I stood transfixed by the live footage of children being marched out of their school into the woods for safety. Walking in a single-file line with their hands on one another's shoulders, they looked so small, and so scared. As a mother, I wanted to fly through the TV screen to put my arms around them and protect them. It was a moment of instant heartbreak, made worse when I imagined what the parents of those children must be feeling as they raced to the school, not knowing whether their child was dead or alive—or worse, discovering that their child had been murdered in the sanctity of their elementary school.

I thought of my own kids. How, on their first days of kindergarten, they'd seemed entirely too young, too small, and too vulnerable to go out into the world on their own. Even now, as teens and young adults, they still seemed like babies to me.

"Please, God, don't let this be as bad as it seems," I found myself saying out loud. Although I was raised Catholic, I hadn't prayed in years. Yet here I was, asking for a miracle to minimize the horror I was witnessing.

Devastatingly, what had happened inside the school was far worse than anyone could have imagined. That morning, a twenty-year-old man had used a semiautomatic rifle and two semiautomatic pistols to shoot his way through the locked doors of a small-town elementary school—an iconic representation of the innocence of childhood—and murder six educators and twenty first-graders as they hid in bathrooms and closets.

I pushed the pile of laundry aside and sat down on the bed, dumbstruck. As I covered my face with my hands, I thought of the long list of mass shootings that had happened in recent years. Columbine High School in Colorado. Red Lake High School in Minnesota. Virginia Tech. Fort Hood in Texas. A movie theater in Aurora, Colorado. A Safeway in Tucson, Arizona. An immigrant center in Binghamton, New York. A mall in Omaha, Nebraska. In my mind's eye, I envisioned a map of the United States with dots appearing in the location of each shooting. The image turned my stomach.

I felt overwhelming sadness looking at the faces of the families on my TV. My heart ached for the mothers who were receiving the unthinkable news that their children weren't coming home. But at the same time, I was enraged at the terrible injustice victimizing these children—how the systems and laws that were supposed to protect them had so clearly failed. If our children weren't safe in their schools, they weren't safe anywhere.

I actually said out loud, "Why does this keep happening?"

In that moment, I was disgusted by my own inaction—the complacent way I'd assumed that someone else would do something. After all, I'd seen this play out before—I'd watched the media coverage after every mass shooting. I'd shed tears for the people whose lives were taken by an angry man—later I'd learn that it was nearly *always* a man—with a gun. I'd gotten angry and thought, "Someone should do something!" And then I'd gone back to my normal life, to my job when I was still working, or to my family responsibilities now as a stay-at-home mom.

In my head, I heard only one word in response to my question, and that word was *Enough*. Enough waiting for legislators to pass better gun laws. Enough hoping that things would somehow get better. Enough swallowing my frustration when politicians offered their thoughts and prayers but no action. Enough listening to the talking heads on the news channels calling for more guns and fewer laws. Enough complacency. Enough standing on the sidelines.

I knew I had two choices: move my family of seven to another country with less gun violence, or stay and fight to make this country that I and my family love safer for all of us. I had no idea what that fight would look like, much less how to participate in it. I just knew I had to do *something*.

And while I'd never been directly affected by gun violence, two incidences in particular had made a big impression on me.

The first mass shooting I remember was in 1991 in Killeen, Texas. I was a twenty-year-old college student, living with my parents about two hundred miles away in Plano, Texas. I was home alone on that day, too, watching the news coverage after a man had driven his pickup truck through the window of a Luby's cafeteria and emptied half a dozen high-capacity clips into the crowd of more than one hundred diners. He killed twenty-three people that day and injured twenty-seven others. I tore myself away from the television to answer the phone. When I heard my dad's voice, I burst into tears.

Another mass shooting had recently hit closer to home. In 2012, a shooter dressed in tactical gear set off tear gas grenades inside a movie theater in Aurora, Colorado, during a midnight premiere of *The Dark Knight Rises*. As moviegoers scrambled to get out of the theater, the shooter used a semiautomatic rifle to kill twelve of them and wound fifty-eight others.

The night after that shooting, my daughter and stepdaughter had plans to take my son, Sam, to see the same movie in our hometown of Zionsville, Indiana. Even though Sam wasn't yet thirteen at the time and the movie was rated PG-13, I agreed—he had always loved all things Batman. Just before they left, more news about the shooting came on the TV in our kitchen. I hurried over to the remote to turn it off, but it was too late. Sam had heard the details and was visibly shaken.

My youngest had always been an anxious kid. He never wanted to stay home alone. He often slept on the floor of our bedroom because of bad dreams. He generally ended up in the nurse's office with a stomachache the day of a big test. He had even developed anxiety attacks, which a therapist was helping him with. Things had just started getting a little better when the Aurora shooting happened.

That night in the kitchen, I put my hands on his shoulders, looked him in the eyes, and told him, "It's okay. You're safe. It won't happen here." Sam looked dubious, but he and his sisters were already on the way out the door, and he really did want to see the new Batman movie. So they went.

I didn't think about it again until they all came home, a lot earlier than expected. The girls told me how as soon as they sat down, Sam had gotten upset, saying he was afraid a man with a gun was going to burst in at any moment. When he started crying, they took him into the lobby to try to soothe him, but ultimately they decided to leave.

In the months that followed, Sam had nightmares and anxiety attacks and slept on my bedroom floor most nights.

Now, only months later, I dreaded having to tell Sam about Sandy Hook, particularly because it had happened in a school just like the one he attended. We could avoid movie theaters, but there was no getting around going to school. I spent most of the rest of that Friday thinking primarily about how I would break the news to Sam. Too upset myself to be a steadying force for him, I decided to put off telling him a little longer.

On Saturday morning I tried to balance myself by going to yoga. I didn't want to go. I'm not really the yoga type—my mind is always going, and I'm the first to admit that I can be controlling and tense. On a good day, I dreaded going to class a bit; on this day, the thought of it made me sweat. But I went anyway. Still shattered by the thought of the grief-stricken parents in Connecticut who'd lost their babies in such a brutal, horrifying way, I willed myself into the car to attend class.

Once at class, I waited for a sense of serenity to settle in. I sat in hero pose. I listened to my classmates' whispery ujjayi breaths. But despite my best efforts, I did not feel calm. All I could think of were the pundits I'd seen that morning offering condolences without calling for any changes. To be honest, all I felt was pissed off.

I thought of my prayer the day before, and I knew that my prayer had to involve more than just thoughts. It needed to include action. I had to *do something,* and that something wasn't yoga. So I jumped up, rolled up my mat, and bolted out the door.

Once home, I didn't even take my coat off. I grabbed my laptop and opened it on the kitchen counter while my husband John and the kids milled around, finishing breakfast and piling plates in the sink. I went online to search for support. I thought there had to be some kind of organization already in existence—like a Mothers Against Drunk Driving for gun violence prevention. But all I found were small state organizations working on local gun violence issues—which had made important strides but didn't add up to the nationwide grassroots army I was envisioning—and a handful of think tanks in Washington, DC, most of which were run and staffed by men.

I knew I wanted to be in the company of other women who were connected to the heart of the issue—the fact that more guns and fewer gun laws meant less safe kids. I could sense that moms were the moral and emotional counterbalance to the gun lobby's bluster and posturing. While it was true that the National Rifle Association was incredibly powerful and had covered a lot of ground by making gun extremists afraid their guns would be taken away from them, American mothers—especially now, in the wake of Sandy Hook—were afraid their *children* would be taken away from them. If that wasn't the type of threat that would spark a mama bear mentality, I didn't know what would.

Women are superheroes every day in their families and communities. I wanted to bring that mentality to this fight.

After looking online for almost an hour, I decided to make my own Facebook page called One Million Moms for Gun Control (we would later change our name to Moms Demand Action for Gun Sense in America) in an attempt to start an online conversation with other moms who were feeling the same way I was.

At that time, I had seventy-five Facebook friends and an inactive Twitter handle. I was not a social media phenom.

"You sure you want to do that?" John asked.

"It's just a Facebook page," I said. "Not a big deal."

Then I typed the words that would change my life—and create the nation's first and largest grassroots counterweight to the gun lobby (words that I hope will continue to impact the story of America's gun violence crisis):

> This site is dedicated to action on gun control—not just dialogue about anti-gun violence. Change will require action by angry Americans outside of Washington, DC. We need to organize a Million Mom March in 2013. Join us—we will need strength in numbers against a resourceful, powerful, and intransigent gun lobby.
>
> I started this page because, as a mom, I can no longer sit on the sidelines. I am too sad and too angry. Don't let anyone tell you we can't talk about this tragedy now—

they said the same after Virginia Tech, Gabby Giffords, and Aurora. The time is now.

I wrote this without knowing that there had been a Million Mom March in 2000 on the National Mall calling for gun reform after the 1999 Columbine shooting. I didn't know anything about state or federal gun laws. And I didn't know that a network of gun extremists lay in wait to attack anyone who dared to change the status quo on guns. In retrospect, that one action probably had such a big effect because I had no idea what I was getting myself into. Because I was so afraid for my children's safety, I was fearless in raising my hand to become the tip of the spear.

The Likes on my Facebook post started coming in instantly. So did the messages. I heard from a mother in North Carolina who had resisted the urge to run to school to check on her eight-year-old after hearing the news reports—in just a few weeks she would become the state chapter leader of Moms Demand Action. I also heard from a mother in Houston who'd been out getting her family's Christmas tree when she heard about Sandy Hook—she would go on to create a campaign in her state that persuaded hundreds of Texas companies to put up signage prohibiting open carry on their premises. And I heard from a mother in Silicon Valley who'd walked out of her job when she heard the news—she would never return to that job and became a full-time volunteer instead.

Women everywhere were asking how they could join my organization, and I didn't even realize I'd started one.

As I stood at my kitchen counter, hearing the notifications on my phone and computer ding incessantly, a memory popped into my head. The month before, John and I had been on vacation in Arizona and had our horoscopes read just for fun.

"I see you having a huge following," the astrologer had told me. "And I see you changing laws—not as a politician, but as more of an activist. You're leading thousands of people from your kitchen table."

At the time, I'd shot John a skeptical look. *Me, an activist? Whatever you say, lady.* I was a former public relations executive and a mom who had never signed a petition or been to a single rally.

"You have some soul sisters," she continued. "They'll help you get this done."

I was dubious. But that phrase—*soul sisters*—that's what popped into my head as the messages from moms all over the country poured in. These were my soul sisters, finding me—and in my kitchen, no less!

As the hours went by, I began to feel in my gut that I'd tapped into something powerful. I knew moms would respond to a clarion call for the safety of kids everywhere, but this was more intense than that. This was the unwavering power of a million moms' hearts all channeling their love, their rage, and their strength into something momentous. As one woman after another reached out to me, I had the strange sensation that I was watching a story unfold that I already knew the ending to.

I knew we'd caught lightning in a bottle, but I couldn't stop to contemplate it—I was too busy responding to each and every message personally.

A frenzy of activity kept my emotions at bay the rest of that day and most of the next. I kept the TV off so we didn't have a repeat of the situation after Aurora when Sam accidentally found out about the movie theater shooting. But as Sunday night approached, I knew I had to tell Sam about what had happened.

I wanted him to hear about the shooting from me, not from a friend or his teachers at school, because I was certain it would send him into another anxiety spiral. As I went into his room to deliver the news, I braced myself for the worst—a meltdown or tears. I made a mental note to call his therapist the next day to help us deal with the fallout. Instead, when I told Sam what had happened, he looked at me and said, "I understand. That's just what happens in America, Mom." Then he casually returned to playing a video game.

I was gutted by how quickly Sam had transformed from a boy who was traumatized by the news of a mass shooting to one who was barely fazed by it. It wasn't an indication of his getting older; it spoke to the horrifying effect that the increasing frequency of these events was having on our entire nation.

Sam's nonreaction showed me in a way that hadn't fully crystallized before just how helpless we all felt the situation had become. It deepened my resolve to lock arms with other women and mothers to do something about gun violence and to fix our broken country. I knew I couldn't live like this anymore. And I sure as hell wasn't going to let my children die like this!

When the kids were back at school the next day, I started looking more deeply into gun violence in America, and I was shocked by what I learned. Mass shootings had gotten my attention and sparked my motivation, but I hadn't realized the extent to which they are just tiny drops in an ocean of gun deaths every year.

I had no idea that an average of ninety-six Americans are killed by guns every *day*—for a total of around thirty-five thousand people every year—and that seven of the people who die each day are children. And I had no idea that, at that time, there was nearly one gun in the United States for every person (that number has since risen so that now there are more guns than people in the US). I learned that America's rate of gun homicides is twenty-five times higher than the rate in other high-income countries. I saw so clearly that we had given the gun lobby's experiment of "more guns and too few gun laws" plenty of chances to prove itself, and it had failed. Miserably.

I also saw that gun safety is an issue that directly affects women—and not just those of us who are mothers wanting to protect our kids. Even though school shootings and other shootings in public places are the most likely to make the news, shootings related to domestic or family violence happen every single day with little media attention. Each month, fifty American women are shot and killed by a domestic abuser. That's more than one every day.

I knew I'd seen only the tip of the iceberg. I also knew that a mom fighting to protect her children was way more powerful than a gun lobbyist fighting to protect gun manufacturers' profits. Looking at the statistics helped me understand that I didn't want to just march, rally, and protest. I wanted to bring together a badass group of women who could go toe to toe with gun lobbyists in every city and state. I wanted to raise an army of tough mothers.

393,000,000 The number of guns owned by American civilians in 2017[1]

329,905,500 The US population as of October 31, 2018[2]

1.2 The number of guns for each and every American

270,000,000 The number of guns owned by American civilians in 2007[3]

25 The number of countries with the next highest rates of gun ownership you'd have to combine to reach the total of American guns in circulation

33,130 The number of Americans who died in gun-related deaths in 2014[4]

36,252 The number of Americans who died in gun-related deaths in 2015[5]

38,658 The number of Americans who died in gun-related deaths in 2016[6]

96 The average number of Americans killed by guns every day[7]

8,300 The number of American kids who are sent to the hospital each year with a gunshot wound[8]

10x How much more likely it is that a black American will be killed by a gun than a white American.[9]

15x How much more likely it is that a black child in America will be killed by a gun than a white child[10]

50 The number of American women shot to death by intimate partners each month[11]

25x How much higher the rate of gun homicide (not including suicide) in America is than in other high-income countries[12]

82% The percentage of worldwide gun deaths that happen in the United States[13]

4.28% The percentage of the worldwide population that resides in the United States[14]

Since that fateful day in my kitchen in 2012, Moms Demand Action has grown to nearly six million supporters and hundreds of thousands of active volunteers, and our numbers are increasing all the time. We have become the David to the NRA's Goliath, despite its hundred-year head start, its long-standing relationships with politicians, and its deep pockets. And despite the unevenness of our match-up, we are winning.

In dozens of states, we've defeated permitless carry, proposals to allow guns in K-12 schools, and bills that would force colleges to allow guns on college campuses. We've helped pass eleven red flag laws—eight of them since the 2018 shooting at the Marjory Stoneman Douglas High School in Parkland, Florida. (Red flag laws provide a legal means to temporarily remove guns from people who are demonstrable threats to themselves or to others; only two existed before our organization was started.)

If you're reading these accomplishments and feeling surprised, it's no accident—our losses get a lot more attention in the media than our wins. After all, the old adage about the news says, "If it bleeds, it leads." And our losses, honestly, mean that more people will die. When we help beat back a bad gun bill—which we do hundreds of times every year—it doesn't get covered because it doesn't have the drama that draws attention.

Despite the quiet ups and the overexposed downs of the gun violence prevention movement, we're making important strides. But our work will never be done. That's a feeling most moms are already familiar with. After all, being a mom is the one job you never clock out of. But somehow, even though there's no way to prepare yourself for the demands of motherhood, you just do it. This is what gave me the courage (or stupidity?) to start that Facebook page. I didn't know what would come of it; I just knew I had to do it.

That *knowing* is maybe something you've felt before, too—a calling to take action, even though you have no idea what that action might be. I've learned that those moments of knowing are gifts. No matter what horrible event led to them, those flashes of insight have so much power in them—and if you act on them, they'll lead you exactly where you need to go. You'll find your soul sisters, and together you'll move mountains.

If this feels daunting, I get it. As a mom, you may feel too overwhelmed by keeping up with your family responsibilities or the deliverables at work to even think about getting involved in such a thing, much less actually creating change. But everything you've done and felt as a mom gives you enormous and uniquely powerful strength.

After all, activism equals organizing, and if there's one thing moms know how to do, it's organize. Moms also have extremely well-honed multitasking skills. We're used to doing all of the jobs, from scheduling and hosting family events to advocating on behalf of our kids to putting our foot down when someone's out of line—all things that translate directly to advocacy. And we have numbers on our side—there are eighty million moms in the United States alone. Moms make miracles happen in their households every day. But if we unite to work together? We're unstoppable.

In a time when so many things divide Americans, the issues that speak to moms cut across party lines, as evidenced by the fact that our membership comes from both red and blue states. Of course, it also includes women who aren't moms—and men. Nothing makes me smile more than seeing a man in one of our red Moms Demand Action T-shirts. Since the earliest days, we have said that Moms Demand Action is for "mothers and others"—but to be clear, *women* are taking the lead.

Because for too long, women have been asked to make the food, set up chairs, and own the menial tasks of advocacy while men set the strategy and bask in the spotlight. Women have done all the work, and men have gotten the credit. It's high time we changed that!

NRA members love to use intimidation as a weapon—later in this book you'll hear about the threats I've received and the men with semiautomatic rifles who show up at Moms Demand Action meetings—but here's a truth for you: moms are scarier than gun lobbyists. Gun lobbyists hide behind anonymity, legislative bureaucracy, and stacks of money. Meanwhile, moms already go into battle every day for the people they love. Activism just channels that warrior energy into a different arena.

Fight Like a Mother is part manifesto, part memoir, and part manual, and I wrote it because I often wished I'd had a guidebook to help me find my way in the years since I first created that Facebook page. Since that fateful day, I and the army of volunteers who fuel Moms Demand Action have distilled eleven principles—mantras if you will—that guide our actions, help us stay on track, and keep us motivated. In this book I'll walk you through each of them so you can take what we've learned and put it to use fighting for the things that matter to you, whether that's commonsense gun laws, reproductive rights, environmental protections, education reform, or whatever else gives you that feeling that it's time to act.

I'll also share a behind-the-scenes look at how Moms Demand Action has become the gun lobby's worst nightmare—you won't believe some of the things that have happened!—but my sincerest wish is that this book will inspire you to get out there, raise some hell, and do some good. It's time to fight like a mother!

1

Use MOMentum

During the first few days after Sandy Hook, everyone was waiting for the NRA to issue a statement. Yet NRA leaders stayed strangely silent. I, along with so many of the other women I was now connected to as my Facebook page grew by hundreds of people a day, took the organization's delay as a sign that it "got it"—that this shooting had gone too far and it would finally realize it was time to make it harder for dangerous people to gain access to guns. It seems so naive, even laughable, that we thought this—especially knowing what I know now about how there appears to be no bottom to the NRA leaders' deranged commitment to removing all legal restrictions to gun ownership. But that's why, when the NRA announced a press conference for December 21, 2012—a full week after the massacre—I couldn't help but feel hopeful and even a little excited to go to the broadcast studio in downtown Indianapolis to provide an on-camera reaction to its statement on MSNBC.

As I sat off camera, listening to NRA executive vice president and CEO Wayne LaPierre speak through my earpiece, I was incredulous. First, he blamed music videos, movies, video games, and the media for school shootings. Next, he called for creating a national database of the mentally ill. Then he uttered the infamous line, "The only thing that stops a bad guy with a gun is a good guy with a gun." And he ended by proposing a nationwide effort to put armed officers in every school and said the NRA would provide the training, free, as a gift to the country.

By the time he was done talking, I had gone from shocked to *pissed*. Which is why, when a reporter from *USA Today* got ahold of me on my phone minutes later, I desperately wanted to say something that was going to get heard; something that reflected how, as a mother, I was offended and in full-on mama bear mode; something that essentially said, *How dare you?* And so I told the reporter, "[The NRA is] about to see a tsunami of eighty-four million angry moms coming out at them. Angry moms like they have never seen before."

Those words came out of me in a moment of maternal fire, and ever since, they've served as a beacon for Moms Demand Action. Granted, my Facebook page was less than a week old, but I knew that American moms would not let this stand. That outrage has been instrumental in helping us grow quickly and stay strong.

Society frowns on angry women—we're often described as being shrill or unhinged; we're called harpies, bitches, and worse when we let our fury show. But those big emotions that offend so many are key ingredients for transformation. As a mom, there is likely always some kind of injustice in the world making you feel rage or anguish or both. Whatever emotion is pulling at your heartstrings, go with it. Those intense feelings aren't meant to torture you, or to make you feel disempowered, but exactly the opposite. Any heartache you may be feeling about where the world is headed means one thing: it's time to go from outraged to engaged.

All you have to do is decide to heed the call—you don't need training, prior experience, or even a lot of time. So many of the active volunteers of Moms Demand Action, myself included, are accidental activists—we were never particularly politically active, and we certainly never imagined we'd become leaders in the gun safety movement. In fact, very few of us would have said we had the extra bandwidth to become activists. But we all felt called to action by a moment that made us realize we could no longer stand on the sidelines.

The volunteers who've joined Moms Demand Action have each had their own moment that spurred them into action: whether it was hearing about yet another mass shooting, or losing a friend or family member to gun violence, or sending their kids to school and finding out that lockdown drills have become a routine part of an American education. Most mothers cannot fathom that their kids—even preschoolers and kindergartners—will regularly spend part of their school day rehearsing for the possibility that someone with a gun will come into their school and murder as many people as possible.

You may not see yourself as an agent of change. (Yet.) After all, you're probably plenty busy taking care of your kids and making a living. You might think you don't have the time, energy, or guts to be an activist. Well, I have two words for you (and I say them with love): *Stop that!* You have so much potential to effect change—more than you know.

I'm not the only one who thinks so. I had the great pleasure and honor of interviewing Congresswoman Nancy Pelosi for this book, and here's what she had to say to me about the power of moms: "Being a mom, what are you? You're a diplomat, focused on interpersonal relationships. You're a chef. You're a chauffeur. You're a problem solver. You're a nurse. You're a health-care provider. You do so much, and that's just with the children, not to mention the other aspects of family. And moms bringing those collective skills to an issue make us unstoppable. Never bet against moms—we are organized, mobilized, and determined to advocate for our children's safety."

The power of mothers to effect change is not a new phenomenon: women have been the secret sauce in the progress we've made on many social issues throughout history. Just look at Prohibition. In the 1800s, chronic drinking in the United States had contributed to many social problems, including the abuse of women and children. Eventually, women began to organize, which gave rise to the Woman's Christian Temperance Union. Because sobriety was considered a Christian value, women—then the religious standard-bearers of American families—were allowed to be on the front lines of the war to eradicate alcohol. Women never looked back and continued fighting political battles in America to end child labor, expand voting rights and civil rights, stop drunk driving—all the way up to exposing the water crisis in Flint, Michigan. It's almost always women who are leading the charge for social change, and gun violence prevention is no different.

What Is MOMentum?

We use the term *MOMentum* a lot in our planning meetings and everyday conversations with each other, because it reminds us that Moms Demand Action is an organization of moms and helps us remember to be loud and proud about that fact. Because the truth is, no matter what race or socioeconomic class we are, we women—especially those of us who are middle-aged (like I am) or older—are not paid much attention in this country, despite the fact that we do so much of the heavy lifting. Seeking to build MOMentum is about giving ourselves the chance to lead and bring about change on the issues that matter to us. It's about always remembering that although we are moms, we are also activists—and those two roles are not conflicting. Rather, they each strengthen the other.

You don't need to chain yourself to a fence outside the White House or be handcuffed by security guards to be an effective activist. These days, you don't even need to leave your house. Many of our volunteers have only a few minutes here or there in a typical week to devote to the cause. Still, it matters when they wear their Moms Demand Action T-shirts while running errands, or send emails during lunch breaks, or fire off a tweet before bed. From the very beginning, we've advocated something we call *naptivism*—a term inspired by a volunteer who made a video while her child was taking a nap to show how to call your member of Congress and then posted it to social media. Some activism is always better than none. Every action, no matter how small, is like drips on a rock—over time, they can carve a canyon through even the thickest, most immovable layer of rock.

Why Moms Make the Best Activists on Earth

Everyone can get engaged in the world. Even moms. *Especially* moms.

You may not realize just how powerful we moms are. After all, it's an undeniable—and shameful—fact that women hold very few formal positions of power in the United States. At the start of the 2019 legislative session—when we hit our highest numbers ever—we still made up only 28.5 percent of state legislatures[1] and 23.4 percent of Congress.[2] In the business world, women make up only 1 percent of Fortune 1000 CEOs.

Yet there are other, more empowering numbers that are too often overlooked. Namely, that women comprise the majority of the voting population. On top of that, we make 80 percent of the spending decisions for our families. Politicians and companies care very much about what we have to say. And when we band together, we absolutely have enough influence as a voting bloc and an economic force to create change.

Another thing that stands in the way of seeing just how strong we are is the stereotype that moms are frazzled and need several glasses of wine just to recuperate from all that cleaning, errand running, homework wrangling, and schedule managing. The NRA has latched on to this perception and often tries to insult Moms Demand Action members by saying that we like to drink boxed wine in our driveways—a stereotype that began when NRA spokesperson Dana Loesch said that I seemed like a "lonely woman who sits in her driveway drinking boxed wine" in a video she made in 2014.[3] Don't get me wrong—some of us might enjoy a sauvignon blanc from time to time, but this does not define us. (And I'm pretty sure many NRA members are boozing in their backyards.) The fact is, we have to own our mom status and not let it be seen as a weakness. If we don't claim our motherhood as a tool, it will be used against us as a weapon.

Trust me, I get that it's easy to feel overwhelmed by the craziness of raising a family and probably also earning a living—after all, I've got five kids and have worked a variety of jobs while raising them. But let's reframe our ability to manage all the things we do as the perfect qualifications for moving mountains.

Moms are formidable. Don't believe me? Think of all the things you've done since the moment you became a mom: giving birth, for one (no matter how that baby came out, how much more powerful can you get than creating a new life?); foregoing sleep; catching every virus your kid brought home from daycare or playgroup; advocating for your kids at school, at the pediatrician's office, or maybe even within your own family; developing the ability to manage multiple people's seemingly incompatible schedules and needs; continually growing your capacity to love other human beings beyond what you ever thought possible; honing your patience; and building physical, mental, and emotional resilience.

And those qualifications are just the beginning of what makes you, as a mom, such a force.

A mother's love is fierce. It's in our nature to protect our children, and that instinct is so strong that we would put ourselves in front of a speeding train to save a child. We are true warriors when it comes to our kids. And you don't even have to try to summon this courage; it's all instinct. Agatha Christie wrote, "A mother's love for her child is like nothing else in the world. It knows no law, no pity. It dares all things and crushes down remorselessly all that stands in its path." The next time you're doubting your ability to disrupt the status quo, remember that—especially the "crushing down remorselessly" part.

For mothers, the thought of losing a child is unbearable. And while suffering such a loss is terrible to contemplate, it's also liberating. It empowers you to protect your kids as if you have nothing to lose. Because if you lost your kids, you would feel as though you'd lost everything.

Of course, moms are also forces of love—the ones who kiss the boo-boos, dry the tears, and teach the emotional lessons. That makes us moms an important voice of morality and compassion—not just in our own families, but in society at large. It also makes us the perfect counterbalance to the posturing and intimidation of the NRA leadership and their lobbyists. After all, our current gun laws are a textbook representation of masculinity gone haywire, and for too long, men have dominated the discussion about guns. Moms are the yin to toxic masculinity's yang.

And it's not just your kids that your heart guides you to protect. As Hillary Clinton said, there's no such thing as other people's children. Becoming a mother makes you realize that you're a caretaker not just for your own kids, but for everyone else's too. You understand that it's part of your role to make the world a better place for everyone; it's a moral obligation that you feel not so much as a duty but as a simple fact of life.

I've seen this natural tendency to support others every day, when our volunteers come together to support those whose loved ones have been taken by gun violence. In 2018 in Austin, Texas, one of our moms, Diana Earl, was facing the emotional gauntlet of attending the trial of the man who had shot and killed her only child, Dedrick. The defense attorneys were trying to paint her son as a thug who had brought his own death upon himself. Members of her Moms Demand Action chapter, many of whom had never met her, organized themselves so that at least a half-dozen of them were at the trial every day, so that she would know that someone was always with her and visibly supporting her in the courtroom.

We've supported each other through illness. Another Texas mom, Catherine Nance, joined Moms Demand Action during the campaign to fight a bill that would allow guns on college campuses in that state. She was an adjunct professor and mom to three little kids and came to every hearing despite having some health issues that she thought were related to her last C-section. When she was diagnosed with stage-four colon cancer, our volunteers cooked her family food, sat with her during her hospital visits, and ripped out soggy drywall in her home after Hurricane Harvey. Catherine remained devoted to the cause to the end, even going out wearing a wig to canvass for a Moms Demand Action volunteer who was running for local office. These women who never knew each other before became as close as family.

After the 2016 mass shooting at Pulse, a gay nightclub in Orlando, Florida, protestors showed up at an Orlando vigil to harass attendees, including one of our volunteers, Wayne McNeil, a gun violence survivor and member of the LGBTQ community. Moms Demand Action volunteers simply walked over to where all the haters were gathered and hid them from vigil attendees with a huge Disarm Hate sign. Wayne later said, "That we could support each other in a moment of unspeakable grief speaks volumes to what Moms Demand Action really stands for in America. People finding the strength to lift each other up with love in the midst of so much violence and hate is what makes us special. It is what we do every day."

This drive to protect other people—even complete strangers—is what makes moms relentless. My husband told me after I started Moms Demand Action that even if he found out he had a terminal illness, he wouldn't spend as much time trying to cure it as I was spending trying to eradicate gun violence. When women are passionate about an issue, they do not take "no" for an answer. That passion is our power; we care so much that we become unstoppable.

So it's not that we can be effective activists *despite* being mothers; we wield power and have access to unfathomable strength *because* we are mothers. It's time for us to wear our motherhood with pride, and beyond that, to use it to our advantage.

Don't Hide the Fact That You're a Mom: Own It. Use It.

When I first started working in corporate America more than twenty-five years ago, motherhood still had a big stigma attached to it. In 1999, when I was the first pregnant vice president at the PR firm FleishmanHillard in Kansas City, I got pushback just for wearing maternity clothes. One day I was in an elevator with another woman (a woman!) who ran the office. I was wearing a suit with a swing-style blouse underneath my jacket, because I couldn't button a normal shirt over my belly. She gave me the side-eye and said, "That's an interesting top you're wearing." Pregnancy was seen as an affliction rather than as a part of life. And motherhood was something I tried to hide as much as possible, saying I was sick instead of admitting that I needed to stay home with a sick kid, and never, ever allowing any telltale kid noises to be heard in the background of conference calls I did from home.

It wasn't until I started Moms Demand Action that I fully understood just how much political clout and power motherhood gives us. Clearly, with a name like Moms Demand Action, we fly the mom flag proudly, but that's not the only way we put the fact that we are mothers to good use.

From the very beginning, we've made it plain that we understand that the majority of our volunteers have kids, and we know that kids come first. You can't get on a Moms Demand Action conference call without hearing at least one child in the background, and kids are always welcome—to chapter meetings, to advocacy days at statehouses, to marches.

We've also co-opted the mainstays of motherhood to remind politicians and influencers that we're watching them, and we expect them to be on their best behavior. After all, they have mothers, too—and they're probably really scared of them. We've brought homemade cookies to meetings with lawmakers, we've communicated with them through crafts we create with our kids, and we've even sent them valentines.

The valentines started the first Valentine's Day of our existence, just two months after Sandy Hook. In a tactic I didn't fully think through, Moms Demand Action asked Americans to mail valentines to their lawmakers. We had people send the cards to a P.O. box in Washington, DC—where I then had to fly in order to sort them and deliver them. And because you can't just take mail through security at the Capitol, I and other volunteers had to smuggle them into Congress under our coats and in our purses (did I mention that moms are also ingenious?).

But all those logistical hurdles were worth it, as delivering those valentines is how I got to meet Joe Donnelly, my home-state senator from Indiana. He told me that an assault weapons ban was a "nonstarter" but that he supported background checks. That was information I was glad to have. That was also when I first met Senator Chris Murphy from Connecticut. As if I needed another reason to like him (he is a staunch gun-sense supporter, particularly because Sandy Hook is in his home state), he had his staff make valentines for Moms Demand Action volunteers. Some may have seen our valentines stunt as juvenile or naive, but would the traditional method of writing letters have made as much of an impact? I don't think so.

We've even used a hallmark of childhood—the lemonade stand—to sway lawmakers. After Congress failed in 2013 to pass the Manchin-Toomey amendment to expand background checks on gun sales, moms were shocked and disappointed at the Senate's failure to act, but we were not deterred. In response we launched a lemonade stand offensive that August, when Congress was on recess. Proclaiming "When Congress gives moms lemons, we make lemonade!," we recruited our kids to help us host lemonade stands in front of legislators' offices with three main goals: show lawmakers that we hadn't forgotten about the issue; build awareness about the issue and our organization; and raise small amounts of money that could be used locally.

We even had a lemonade stand on Capitol Hill after Congress was back in session. Notable lawmakers like Senator Tim Kaine from Virginia (and later vice presidential candidate), Senator Chuck Schumer from New York, and the entire Connecticut congressional contingent bought lemonade from us—and of course we took pictures and posted them on social media.

Along the way, we've taken our kids with us when it makes sense—sometimes out of necessity, sometimes as a strategy, and sometimes as both. One of our first and most effective tactics to encourage resistant lawmakers to meet with us is to hold "stroller jams." Stroller jams originated in Maryland, where then-governor Martin O'Malley introduced background check legislation after Congress failed to pass Manchin-Toomey. During the run-up to the vote, our volunteers met with lawmakers and took their kids along. With all the strollers and car seats, the halls of the Maryland statehouse were packed. As a result, lawmakers didn't have any room to maneuver past us; they had to stop and talk to us.

It was a crystallizing moment. After the Maryland bill passed, Governor O'Malley thanked our organization publicly, which gave us a lot of credibility and put us on the map as moms and as a political force. Now we use stroller jams in the halls of Congress, at state legislatures, and at the in-district offices of congressional representatives—even on public transportation.

Through these tactics and more, we've seen that all the stereotypes of moms add up to a powerful brand (something I talk more about in Chapter 7). This is a tool you can use to great effect, and it's certainly not something to hide.

If You Can't Walk Through the Door, Go Through the Window

Moms may not have a lot of direct access to power in government or from within corporations, but our purchasing power is an incredible tool. How you vote with your dollars sends an important message about what you believe in, and about what the companies who want to sell things to you ought to believe in too, if they want your money.

Moms Demand Action has had a lot of success in targeting companies to change their gun policies. The first company we set our sights on was Starbucks. In June 2013, just six months after starting Moms Demand Action, I saw on the news that Starbucks was going to prohibit smoking and e-cigarettes within twenty-five feet of its stores—no matter what the state laws said about regulating smoking in public places.

But at the time, Starbucks allowed open carry inside its stores. "Open carry" means that someone can carry a handgun or even a long gun in plain view, unlike "concealed carry," which means that a gun cannot be seen by a bystander. In many states, open carry is unregulated, and unbelievably, forty-five states allow it.

Even more shocking, earlier that year, on February 2 (get it? the second day of the second month—to honor the Second Amendment), gun extremists organized a day where they encouraged people in open carry states to take their handguns and long guns to Starbucks. It was sickening to watch. And it was disturbing to think that this company's stores, where moms spend so much time and money—often with their kids—were becoming hangouts for gun extremists.

So when I saw the news about Starbucks banning cigarettes regardless of state law, I thought, "Wait a minute, they're going to continue to allow guns *in* their stores, but they won't allow cigarettes within twenty-five feet?" It was outrageous to me that Starbucks didn't understand that secondhand bullets are more dangerous than secondhand smoke.

So I called the Starbucks corporate office and asked whether company leaders were still going to allow open carry. They told me they were. Not only were moms and their kids going to continue to be exposed to gun extremism in a business that was such a big part of their lives, but open carry supporters were being sent the signal that Starbucks CEO Howard Schultz fully supported them. Starbucks was starting to feel like enemy territory to a huge segment of its patrons.

I considered calling on our volunteers to boycott Starbucks, but remember, Moms Demand Action was only six months old. We didn't yet have the numbers to pull off a full-scale rejection of such a beloved and powerful brand. But we clearly needed to change the narrative so that Starbucks and the rest of the country could see that someone carrying an AR-15 openly while they ordered a latte was being considered safer than cigarette smoke.

We settled on a "momcott" that we called Skip Starbucks Saturday. Withholding our patronage was just one objective; just as important, we wanted to raise awareness of how pervasive open carry had become by making the images of open carry inside Starbucks go viral on social media. We put out a series of press releases and social media posts encouraging moms to post pictures of themselves drinking coffee at home or a Starbucks competitor and use the hashtag #SkipStarbucksSaturday.

The more we pushed against open carry inside Starbucks, the more gun extremists pushed back. They showed up at Starbucks from Cincinnati to San Antonio—even near Sandy Hook Elementary School, where the horrific mass shooting had happened less than a year earlier. Many of the images from these visits went viral, including one of a man and a woman holding Frappuccinos inside the store with AR-15s strapped to their bodies. In Sioux Falls, South Dakota, more than a dozen gun extremists went to Starbucks armed to the teeth. The patrons and staff were terrified, but management said there was nothing they could do because company policy allowed it.

Our "momcott" had been going strong for about four months when one sleepless September night, I received a Facebook message at 1 a.m. from someone I didn't know that said, "Congratulations, you won." Starbucks had sent out a late notice to its employees that guns—all guns, not just the open carry of guns—were no longer welcome inside Starbucks stores. This was an unbelievable victory: one of the most powerful CEOs at one of the most prestigious brands had listened to a group of angry moms.

What was even more exciting was that this wasn't public knowledge yet; I found out about it from a Starbucks employee. My PR instincts started firing, and I saw a chance to take control of the narrative and make this story about the power of moms to influence corporate policy and social change. Since I was awake anyway, I wrote a press release. I texted around to find a Moms Demand Action volunteer who was both a graphic designer and also awake. No surprise, I found one—and we posted everything by 3 a.m.

I found out later that Starbucks had written press releases days earlier—that hadn't mentioned Moms Demand Action—but had embargoed the announcement (PR language for asking media outlets not to report on the information until a specific date). Howard Schultz was planning a media blitz to make his announcement the next morning.

I spent the next two days doing interviews and sharing Moms Demand Action's perspective on open carry and corporate responsibility. And all thanks to moms' ingenuity (and, okay, insomnia), we were able to get out in front of a huge story.

I imagine that the Starbucks higher-ups shit a brick when the PR team told them that morning that the cat was already out of the bag and they now had to follow our lead—which, by the way, was the lead of the company's customers. It brought me such glee to see our scrappy organization take on a big, well-oiled machine. It was also very gratifying to take the credit we earned that we likely wouldn't have gotten otherwise.

Ultimately, we helped force cultural change and shift what's considered normal and acceptable. And we garnered even more momentum: once we won this battle, we won more. Later, Moms Demand Action would go on to launch campaigns targeting other businesses that moms patronize about their open carry policies (I share more about those successes in Chapter 5).

And while social media are important components of our strategy—those pictures of people carrying weapons inside Starbucks did as much as or more than our small boycott did to sway public opinion and change Howard Schultz's mind, I'm sure—its root is still the fact that most companies know they need the support of moms to survive.

I got a fair amount of pushback from our volunteers about #SkipStarbucksSaturday. I heard a lot of "Why are we devoting so much time to Starbucks? We need to stay focused on legislation." While their point was valid, you leave too much power on the table when you don't use every tool available to you. Thanks to that campaign, we not only changed Starbucks corporate policy, we also made Americans aware of open carry laws and their inherent danger—something no one was talking about before Moms Demand Action called attention to it.

The fact is: Companies need moms' business. They need our loyalty. And when lawmakers don't protect their constituents, companies now understand that they need to protect their customers.

Let Your Fear Motivate You

Don't worry if you feel like you don't know what you're doing or what to say, or if you're afraid of what others will think if you take a stand. There is no weakness in being afraid. In fact, although it's uncomfortable, fear is a great motivator. Ironically, a huge fear—like the fear that your child will be traumatized or, worse, shot at school, on the sidewalk, or when pulled over by the police—will get you to start doing the smaller things you're scared of, like standing up and speaking out.

The NRA has been using fear to motivate its members to be vocal for decades because those members are (irrationally) afraid something will be taken away from them—their guns, their rights, their privilege, and their power. And with all due respect to the Second Amendment (which, by the way, I fully support), being afraid that your kids will be taken away, or that every American kid's childhood innocence will be taken away, is a way more powerful driver.

One thing moms are not afraid of is the NRA. We've raised teens, after all. If anyone should be afraid of anyone in the gun debate, NRA lobbyists should be afraid of moms. So should the politicians who kowtow to the NRA. After all, a woman brought you into this world, and women will take you out—out of your elected seat in municipalities, statehouses, and Washington, DC.

Magic Happens When You Walk Your Talk

We teach our kids not to be bullied and to stand up for themselves. And when we step up to fight for a cause we believe in, we model that action for kids, which is the most influential way we can impart the things we want our kids to learn. When our kids see us stepping up, we inspire them to take a stand for the things they care about too, whether they do that now or years from now. We all need to see examples of people we identify with doing things that might otherwise seem impossible.

When you become an activist, you model that for your kids.

Something magical also happens when you dare to do things that scare you: you give yourself the opportunity to find the people who feel the same way you do. This is where that uniquely female trait of bonding and supporting one another comes in. One Moms Demand Action volunteer said it best when she wrote on our Facebook page: "Sometimes I feel like I'm out on a high wire when I speak publicly or when I attend an event and armed opposition shows up, but I also have this real sense . . . that Moms are forming a safety net below me. I feel like being myself and speaking our truth is finally safe. Moms charge my batteries so I can go out again and again on the high wire."

The only catch is that your soul sisters won't be able to find you if you never open your mouth and speak—whether it's at a live event, while waiting to pick up your kids at school, or on social media. Getting involved in an issue you care about helps overcome the undeniable fact that motherhood can be isolating—it's easy to feel siloed in your own house. When you join a cause and find the women who feel similarly about an issue you're passionate about, it's such a relief. Especially if you live in a place where your ideas and thoughts are in the minority. Wherever you live, you realize just how many other women have your back, and have it in a big way. That's why our volunteers get tattoos and go on vacations together; an undeniable bond comes from being in the trenches.

You may also be afraid that your family or friends will disapprove if you get involved in a polarizing issue. But the truth is, people will respect you when they see you doing something that's clearly scary. When Sabine Brown worked on starting a Moms Demand Action chapter in Oklahoma, her dad was very upset. He was a gun owner who supported gun rights; even more than that, he was afraid for her safety. But she didn't let his concern and disapproval stop her. She kept inviting him to attend events and kept sharing why she cared so much about gun safety. He went from opposing her to supporting her to showing up at legislative sessions wearing his red Moms Demand Action T-shirt. When you demonstrate your commitment, the naysayers in your life often become supportive—and sometimes they even get involved.

Let the knowledge that you're on the right side of history bolster you to get started, and know that once you do, your people will find you. In the beginning you may only have the courage to show up. But that still counts. And once you get over that initial fear of being seen, it will get much easier to start opening your mouth.

2

Build the Plane as You Fly It

Five and a half weeks after starting the Facebook page that would unleash an army of angry moms on the NRA, I found myself on a stage in front of hundreds of thousands of people. It was January 2013 in Washington, DC, and I don't know if it was the weather or my mortal fear of public speaking, but it felt like the coldest day of my life.

I was sitting between the actress Kathleen Turner and Secretary of Education Arne Duncan, waiting for my turn to address the crowd at the March on Washington for Gun Control. My hands were shaking so badly—again, whether because of the cold or my nerves, I couldn't tell—that I could barely read my notes for the remarks I had prepared. I thought to myself, "If anyone had told me that starting this group would involve public speaking, I never would have done it."

In my former life, I used to call in sick on days I was scheduled to give a presentation. *To my own colleagues.* Walking up to that podium on that frigid January day now seemed like more than I was physically capable of doing. What was I doing here? Who cared what I, a stay-at-home mom from Indiana, had to say about gun laws? It took everything I had to stay where I was and then make my way to the podium when it was my turn to speak.

That certainly wasn't the only time I wondered whether I was doing the right thing. There were many moments in the days and weeks after starting Moms Demand Action when I felt overwhelmed, underprepared, and generally exhausted by the never-ending demands of my new volunteer position. In a matter of days, I had gone from a stay-at-home mom who could come and go as she pleased to being tethered to my phone and computer from the moment I woke up until it was time to go to bed again. In those early days, there were weeks when I and other early volunteers worked fifteen-hour days, seven days a week.

Each night, I slept with my phone under my pillow. I'd reach for it as soon as I opened my eyes and start catching up on the emails and social media posts that had come through during the night. I'd get on my first conference call while I was getting the kids ready for school. In those days, I was on conference calls all the time—while driving the kids to school, at the grocery store, at my kitchen table.

John and I had been married for only five years at the time, but he was a godsend. He had always cooked dinners, but now he started making breakfasts and lunches, too. He was supportive, even when I started traveling for the first time in our marriage—we'd spent pretty much every waking moment together since getting married in 2008. Suddenly, I was regularly flying to Washington, DC, Silicon Valley, and New York City.

One day, when I'd been asked to fly to DC for a White House press conference in twenty-four hours, John looked both sad and angry. "I miss you," he confessed. "I miss the life we had before you started this Facebook page." I felt for him. I missed him, too. But it was almost like I didn't have a choice—Moms Demand Action had taken on a life of its own, and I couldn't let down the volunteers who had already devoted so much time and energy to the cause or the kids whose lives we were trying to protect.

I went from being a helicopter parent to a parent who was barely able to keep track of all of my kids' school and sporting events, let alone attend them. I missed choir concerts and soccer games and awards ceremonies, but my kids never complained—in part because so many times when I couldn't be there, John attended in my place. I felt guilty at times, but I also felt fulfilled, and I knew the work I was doing would ultimately benefit my kids—and their kids.

I couldn't help but throw myself into this work; our momentum built so quickly that I could almost feel my hair blowing back. Even though I didn't know completely what I was doing, every cell in my body told me that if I tried to figure out what I was doing before I did it, I would miss the moment, like a boogie boarder who gets left behind by a wave. It was exhilarating to grow so quickly; it also made me see that I had to let go of my long track record of making plans and following them to the letter, checking off every item on my list along the way. I had to let things be good enough.

You Have to Get Started Before You Think You're Ready

I want all women to get this message, particularly every type-A woman who is prone to cross every t and dot every i before diving in: you can't wait until you, your message, or your circumstances are perfect to start creating the change that you know you're here to play a role in making. I know that all the research, planning, and preparation you typically do before you make a move comes from an honorable place. You want to do a good job, after all, and there is nothing wrong with that. But if you dig a little deeper into what's making you want to be completely buttoned up before you make a move, I think you'll see that what's really at the heart of all that perfectionism is fear: fear of making a mistake, and particularly—God forbid—of making a mistake in public.

Getting started before you're ready requires you to trust in yourself. If you don't, you'll try to tell yourself that you're incapable or unworthy of leading, when in reality, you're more than adequately—dare I say, *perfectly*—qualified. If you wait to be asked to lead or try to create a consensus to move forward, what could have been will pass you by.

You don't have to be impeccable to start something. You just have to feel you're called to do it and believe that your passion, smarts, and fortitude will see you through.

When I started Moms Demand Action, some people told me I was reinventing the wheel. They said I should just join another organization that already existed. Some said I wasn't the right person to lead because I hadn't been personally affected by gun violence. Others said that living in the middle of Indiana would make it too difficult for me to connect with lawmakers and the media.

While I was open to their input, I gut-checked everything they said. And each time, my gut told me they were wrong. When I sat down to think about any objections as to why I shouldn't plow ahead, I realized that angering others or failing was less upsetting than not trying at all. And so I didn't wait for anyone's permission to move forward. I *gave myself* permission—and you can do the same for yourself, too.

I've met other women who are or want to be activists, and I've noticed that a desire to be perfect frequently gets in their way and limits their impact—and that is *not* the way to build trust in yourself. For example, a friend who is a national corporate activist hadn't responded a full day after a company had violated a policy in her platform. I reached out to her to ask why. She said she needed to do research, then her board would review her messaging and materials, and then she'd send out a press release. She expected to have something out by the end of the week. I knew that the moment would have passed by then, but I didn't want to give her a lecture. So I simply asked her why she was doing it that way. She admitted she didn't want to get caught making a mistake.

There have been so many times when pulling something together quickly has helped Moms Demand Action further our cause. For instance, in the weeks leading up to the 2018 midterm elections, I saw a tweet from a Virginia town councilman who said he'd be happy to take out any protestors with his semiautomatic rifle. I re-tweeted it. The next day, the *Washington Post* wrote an article about it, and then he deleted his Twitter account. It didn't get him out of office, sadly, but it at least took away a platform from which he could threaten his constituents and spread divisive and violent rhetoric. If I'd waited and thought about it, the moment would have passed and he'd likely still be broadcasting from his bully pulpit.

There is a healthy balance between being accurate and acting quickly. One doesn't need to cancel out the other. It takes confidence to act even when all your ducks aren't in a perfect row, but once you start doing it, you build the trust in yourself to know that if you make a mistake, you'll figure it out. Waiting until everything is perfect only puts you at the end of a long line of people who have the same idea.

I've spoken to so many women over the years who told me that they had the same idea after the Sandy Hook shooting: that American mothers had to come together to do something. But they couldn't imagine themselves being the answer to the question "Who will lead?"

There was certainly nothing extraordinary about me that made me the right person to help galvanize American moms around the issue of gun safety. I wasn't connected to anyone or anything inside the DC Beltway or even politically active. I wasn't a gun owner or a policy expert. And I'd never been directly affected by gun violence.

I want to acknowledge the privilege of that last point: how lucky I was that gun violence had not hit closer to home for me than Sandy Hook. Many people of color don't have that luxury. Too many Americans have never had the privilege of standing on the sidelines, or even feeling safe in this country. Until Sandy Hook, I'd never worried that gun violence could affect my family. My biggest worries were about my kids making friends and getting good grades. I realize now that I was living in a bubble.

But what I knew for sure was that as a former PR executive with an almost twenty-year career under my belt in branding and marketing, I had a skill set I could use to create a clear message and a sophisticated, compelling brand; and to attract attention to the cause—which turned out to be half the battle in the early days of Moms Demand Action. I had to let that be enough. I knew that Sandy Hook had created an opening for moms, and I wanted to maximize it. So I didn't wait until all of my ducks were in a nice line. I just jumped in.

Be Very Selective About Whose Voice You Let in Your Head

The first year after I set up that Facebook page, there were many times—during moments of frustration, despair, or fatigue—when I thought about giving up. If it weren't for the encouragement and objectivity of the Moms Demand Action volunteers—women I'd met only on the internet but whom I came to trust with my life—I'm not sure I would have had the fortitude to keep going.

And that brings up an important point: you have to be very careful about whose voices you let in your head. If your only sounding board is someone who may have good intentions but just wants you to stay safe, you might get discouraged at the exact moment when you need encouragement to keep going. If you were brand new to being a vegan, you wouldn't want to go to a barbecue festival. Make sure the people you surround yourself with have fortitude to spare for those moments when you might start questioning your path.

Every volunteer I entrusted in those early days treated our organization like a start-up. They made it clear that they understood we were building something of value from scratch. They were willing to put in long hours, overcome constant obstacles, and sacrifice the comfort of their former lives to save the lives of strangers. In return, despite the thousands of miles that separated most of us, we became the closest of friends and allies, united against a formidable opponent, and driven to create a safer, better future for all our kids.

Together, we hung in there through all the obstacles that threatened to dismantle the plane we were building, even after we'd already taken off and were up at ten thousand feet.

Learn What You Don't Know from the People Who Know It

Even though I absolutely believe in building the plane as you fly it, I also think it's a good idea to read a manual now and again. Whether you're starting a movement or getting involved in an existing one (Moms Demand Action is always open to new volunteers!), it's vital to learn what you don't know—as long as you also keep in mind that you'll never know everything. Get up to speed, but don't spend more time than necessary on what's happened so far or else you'll be too busy looking backward to make a difference.

I knew that I was no expert in the history of gun policy when I started that Facebook page. I had a lot of learning to do. We all do. So one of the first things I did was to reach out to every expert I could find. Some I found online, some were recommended to me, and some were people who were already nationally known. Most I didn't know personally. In some cases I cold-called them, and in others the women who were helping me made the connection. It really didn't matter how many degrees of separation stood between us—they were all receptive to my requests for help and graciously gave me access to their guidance and insight.

Given my original desire to join something like Mothers Against Drunk Driving (MADD), I knew I wanted to speak to Debbie Weir, that group's CEO. I didn't know how to reach her, so I called the MADD helpline and was blown away when Debbie herself answered the phone. I told her about my Facebook page and the subsequent outpouring of support, and over the course of the next few weeks she spent hours with me and other volunteers on the phone explaining how MADD had organized to take on the alcohol lobby and the lawmakers beholden to it.

Debbie taught me so much about working with survivors of trauma, how it's a whole different ball game from working with your typical volunteer, and how to help organize them while being respectful of everything they've been through. She was an invaluable resource who became an ally. I'm happy to report that years later, she also became the managing director of Moms Demand Action, and we are so lucky to have her.

Another woman I reached out to right away was Laurie Kappe, a California gun safety lobbyist I had met just a few months before on a yoga retreat. She and I sat next to each other every day at lunch, and she would regale me with stories about legislative campaigns that at the time, frankly, I had no interest in. I was there to seek inner peace, after all! After Sandy Hook, she was one of the first people I reached out to and was hugely knowledgeable and helpful. Laurie counseled me to be moderate in our messaging. She told me to be clear that we are not anti-gun, but rather to focus on finding nonpartisan solutions that could unite people in both blue states and red states.

I also relied on the expertise and knowledge of women who had been doing this work for years, including Donna Dees-Thomases, who had planned the Million Mom March in May 2000 after a 1999 shooting in California, and Nina Vinik at the Joyce Foundation, a Chicago-based public policy foundation.

Because the initial name of the Facebook page was One Million Moms for Gun Control, Donna had already been receiving a lot of inquiries from people thinking our organizations were connected. She was so gracious about sending those folks our way; Moms Demand Action is really built on the shoulders of the work she began, and I am so thankful for her tireless contributions over the decades.

And Nina counseled me to advocate and lobby, not just hold marches and rallies. She connected me to organizers who volunteered their time and expertise to help Moms Demand Action create the organizational structure that has served us so well (more on that in just a moment).

All these women helped us refine our strategy, messaging, and tactics. Along the way, they brought donors and partners into our fold and sent volunteers our way from all corners of the country. The help is there if you are willing to ask for it.

One of the best ways to track down specific experts is online, through Google and LinkedIn. Or crowdsource for ideas on experts via Twitter and Facebook. Once you have names, pick up the phone and start cold-calling. Ask everyone you speak to whether they can recommend three other people you should call. Follow up on every lead because you never know when you'll be connected to someone who can help support your work in significant ways. And always, always thank people with an email after they've given you their time.

You can't afford to be shy. Don't let your inner critic talk you out of making the ask. You may be able to build a rudimentary plane that will at least take off with only limited knowledge, but if you want that plane to really go the distance, you'll need to borrow the expertise of others.

Trust the People Who Want to Help You

In those first weeks after creating the Facebook page, I was getting calls from women (and a few men) who offered their talents pro bono to help me get the new organization off the ground, including trademark lawyers, medical professionals, website designers, graphic designers, social media experts, advertisers, and professional organizers. Without this outpouring of support from total strangers, Moms Demand Action never would have been able to get to the cruising altitude it enjoys today.

This was another important lesson I learned: although you'll have to ask for some of the support you need, much of it will seek you out. You have to be receptive to those unsolicited offers, even if you don't know what to do with them right away.

(In those early days, I put every email I received offering to help in some way in a folder called "volunteer offers" and would go through and respond to those emails when I had time. Now we have a whole structure set up to handle offers of help from potential volunteers—see page 132 for more on that.)

When I was a manager at General Electric, the people who reported to me would tell me that I was a good manager but that I was intense. I would constantly make lists, assign tasks, and then check in with people to see whether they had completed them—I was a taskmaster. I quickly learned that working with volunteers—*especially* when you're working on a cause, such as gun violence, where the volunteers have likely experienced some form of trauma—requires a totally different approach. You can't micromanage. Volunteers give their time out of the goodness and generosity of their hearts. No one is making them carve an additional workweek out of their already busy schedules; they're doing it because they care. In return, they deserve trust, patience, and gratitude.

One early morning, just days after starting Moms Demand Action, I found myself on a conference call with an amazing team of talented women—Harvard-educated pediatricians, a journalist, a professional organizer, a business leader—several of whom were also gun violence survivors. They were looking to me to lead, but they also were clearly more than competent and not in any way junior to me. In fact, they were pretty intimidating. They clearly had their own opinions and ideas about how to make Moms Demand Action successful, and my not listening to their advice would have alienated them—and rightly so. In that moment, I consciously decided to take a leap of faith and trust every single volunteer to do what they did best and to always assume they'd follow through on their commitment. What also helped me make that leap in thinking was the fact that, even if I'd wanted to micromanage, I didn't have the time. It was all I could do to keep track of the big picture.

Resist the Urge to Do It All

Since starting Moms Demand Action, I've seen the same cycle play out over and over again: a high-performing woman steps into a leadership volunteer position and ends up taking on all the work herself. Despite offers of help from other volunteers, she keeps her knowledge close to the vest and holds on to all the responsibility for meeting her deliverables. She creates the strategy, and then she executes on every single tactic. As a result, she feels overextended and stressed out, and eventually she burns out and quits.

There is a tendency among women—and I am totally raising my own hand here—to want to own everything, in part because we're multitaskers, in part because we're not usually well-versed in asking for help, and in part because we're pretty darn competent. We trust ourselves to get the job done and done well, and we don't have the same level of trust in others.

Every month, I get notes from the organizing managers of our state chapters telling me which volunteers could use a thank-you call. Then I call them and thank them from the bottom of my heart for their efforts and their courage. (I have been known to bawl on people's voice mails; but it's the good kind of tears.) Once in 2018, when reading through these notes,

I saw the name of the group leader from Austin, Texas, Nicole Golden, with the comment next to it, "She could use a pep call, she's been a bit overwhelmed lately." Nicole had been with Moms Demand Action since we were still called One Million Moms for Gun Control. She'd been instrumental in getting our Texas chapter up and running, and she'd been working even harder to accommodate the influx of volunteers we got after the Parkland shooting. I made it a point to call her first.

After I thanked her for all the work she'd been doing, she began to cry. Through her tears, she told me, "I get so much satisfaction from this work, and I know I'm good at it, so I do it all myself. I've tried delegating, but it's been discouraging, so I've continued to take on too much."

Nicole had been in her role for five years. She was amazingly effective at advocacy and organizing work. Lately, in addition to all of the legislative meetings and hearings she'd been coordinating in the wake of the May 2018 school shooting in Santa Fe, Texas, Nicole had been sitting with a gun violence survivor every day at trial. It took her getting to the verge of breakdown to see that she hadn't divvied up her work enough. I think we can all relate.

Nicole's also the stay-at-home mom of two kids, who were two and five years old when she first got involved with Moms Demand Action. Her work with us was intertwined not just with her life, but with her identity—she describes the past five years of her life as "raising babies while also growing this baby organization." That personal investment is great, but it's a double-edged sword. In Nicole's case, any time she thought about delegating some of her work, she talked herself out of it.

By the time she and I got on the phone, Nicole was considering dropping out of Moms Demand Action altogether. She confided in me that she'd had a hard time finding someone to take over some of her responsibilities. She also confessed that it felt like she'd been on her phone, messaging other volunteers and scanning social media, seemingly nonstop. Her emotional health was taking a hit, and she and her husband had started arguing.

These are telltale signs of burnout: blurred lines between "work" and "life"; stress that can bloom into anxiety, depression, or both; and strained relationships. As important as the work we're doing is, it's not worth sacrificing your health or your marriage.

During our call, Nicole and I brainstormed ways for her to step back and protect her state of mind by setting up some better boundaries around her social media use. Shortly after we hung up, Nicole transitioned into a spokesperson role, so she's still very much involved with our cause, but not so much with the day-to-day operations. And she's beginning to look for a career that uses the skills she's developed in her work with Moms Demand Action, such as writing and giving speeches and speaking to the media. That's what sharing the load does—it helps you see new opportunities for yourself that you would likely otherwise be too stressed to recognize. Nicole is now putting a résumé together for the first time in twelve years, and she's excited to realize what she's qualified to do. "I might even have to trim my résumé down a little," she told me.

I know how much you're capable of—really, I do. (If you don't believe me, reread the previous chapter.) But it's a whole lot harder to be effective if you're overwhelmed. For the sake of your own sanity and health, your family's sanity and health, and the cause you care so much about, please, resist the urge to do it all yourself. It seems like the right thing to do, but in the end, it can only end up hurting everyone.

Trying to do everything on your own is problematic for several reasons. First and foremost, it rebuffs the people who want to help—and trust me, you are going to need help! It's simply not possible for a movement to rest on the shoulders of one human being. By keeping other people on the sidelines, you diminish your impact and your longevity. It also prevents people from being trained for other leadership positions, which means that ultimately it's a path to burnout. It's also not inclusive: trying to own everything deprives volunteers of feeling like they have skin in the game. It can be demoralizing for them.

I can't stress this enough: in order to be a successful leader you've got to ensure that the workload is shared. Sharing the work in an organized way keeps engagement high and builds a foundation for long-term growth.

Here's how Moms Demand Action approached it: We created a senior team of qualified volunteers who each agreed to take on a specific chunk of work. I asked each of those volunteers for a serious commitment. In return, they got a title and a lot of ownership (I didn't step in and meddle). The positions on our senior team include membership lead, legislative lead, events coordinator, social media lead, and more.

Our senior team had—and still has, to this day—regular team calls to check on progress and to monitor workloads. Those senior team members found volunteers to help them with their work and had their own separate calls with their teams to divide the tasks and determine deadlines.

For example, the woman who served on the senior team as our events lead is the one who decided we would have lemonade stands at the Capitol and in our communities to raise both awareness about the Manchin-Toomey vote and money for our organization. After the senior team signed off on the idea, she worked with volunteers who'd expressed an affinity for event planning to design the lemonade stands, create the messaging and handouts, and plan the calendar for where and when the stands would be set up. Then she worked with the communications senior team lead to make sure we got media attention.

I found that this strategy of giving volunteers a ton of ownership and autonomy over their respective jurisdictions worked incredibly well. Because we were all women on a mission, we worked collaboratively and quickly.

Another piece of my management style that had to change was how I rewarded people. When I was a corporate manager, I could give high performers raises and promotions. But that's generally not possible with volunteers. The new way I recognize a job well done has become one of my very favorite parts of my work with Moms Demand Action: making thank-you calls, like the one I made to Nicole. It's one of the ways this work has changed me most—I love coming from a place of gratitude instead of hierarchy so much that I'll never go back to a more traditional power structure of "do this because I say so." Self-help books and Oprah tell us all the time to focus more on gratitude, and I can tell you now that it truly *does* help you to look for and pay more attention to the good things that are happening in your life. It's even changed my parenting style. Now, I'm much more likely to text my son to thank him for remembering to put his dishes in the dishwasher before he leaves for school than I am to yell at him when he forgets. (And I notice that he's more likely to remember, too.)

Course Correct as You Go

If the thought of making a misstep or an outright mistake gives you agita, gird your loins, because messing up is part of being an activist. Making mistakes is inevitable when you're tackling a thorny issue—there simply isn't one right way to make the world a better place. The good news is that once your plane is up and flying, you can always readjust your flight path.

One of the very first recalibrations I had to make was changing our name. When I started that Facebook group in our kitchen, I winged it—no focus groups, no consultants, and no surveys. Without anyone to help me pressure-test the name, I called it One Million Moms for Gun Control.

It wasn't until I told my daughter Emma, who is gay, about the group I had started that I learned that One Million Moms was a religious organization that had worked to unseat Ellen DeGeneres, who is also gay, as JCPenney's spokesperson. Oops.

Yet I didn't make any official moves to change the name until a couple of weeks later, when I got a call on my cell phone from a Washington, DC, area code. I answered and was shocked to hear the voice of Rep. Carolyn McCarthy on the other end. Known on the Hill as "the gun lady," Representative McCarthy had been inspired to run for Congress in New

York shortly after her husband and son were shot in a mass shooting on a Long Island Rail Road commuter train in 1993.

Once she had me on the phone, Representative McCarthy skipped the pleasantries and got straight to the point: "We need mothers to rally around the issue of gun safety. We've been waiting for the mothers," she said. "But you can't use the words 'gun control.' They're too polarizing. You'll have to change your name."

With the help of a volunteer who was also a trademark lawyer, we changed our name to Moms Demand Action for Gun Sense in America. "Moms demand action" was a phrase we'd chanted during rallies and marches. The upside of that initial stumble was that it gave us an excuse to put out yet another press release advertising our organization.

Mistakes are what you make of them. They can be embarrassing failures that send you into hiding, or they can be minor challenges that help you navigate future obstacles. I've learned that acknowledging mistakes, correcting them quickly, and then meeting with your leadership team to break down what happened and how to avoid it in the future is as effective an approach for volunteer organizations as it is for Fortune 500 companies.

3

Channel Your Inner Badass

Until the day I found myself sitting at a table in the Indiana statehouse facing a panel of hostile, overwhelmingly white, male lawmakers, I really didn't know what I was capable of as an advocate.

By that point I'd done some things that had scared me: I'd spoken at some events (which, as I mentioned, sent me into a cold sweat—I got through them but I certainly didn't kick ass at them . . . yet). I'd been interviewed on TV (staring into a camera lens and rattling off talking points without looking stiff or scared was a skill it would take me a while to master). But I hadn't actually been in the fight in real time. Sitting in the Indiana committee hearing in March 2014, I felt like I'd mistakenly landed in a boxing ring with no gloves and several opponents with bloodthirsty looks in their eyes. It was a lot different from being behind my laptop at my kitchen table.

I was there at the request of our Indiana chapter leader, who wanted a high-profile person to testify against a law that would allow guns in cars in school parking lots. I'd had only twenty-four hours to prepare, so I'd pulled together some thoughts on a note card. I imagined I'd go in and read my testimony and that would be it.

What immediately became clear was that Moms Demand Action had started to be noticed by the opposition, and in order to take us down a peg, this panel of mostly male lawmakers was dead set on humiliating me. While I waited my turn, I sat in the hearing room next to an NRA lobbyist, listening to the other side spout false data from John R. Lott Jr.—an academic researcher and go-to expert of the NRA whose findings had long since been debunked. I could sense this wasn't going to be the simple exercise I'd anticipated, but I was still shocked when, after I read my testimony at my appointed turn, I faced a firing squad of personal and professional questions.

One lawmaker held up a copy of my résumé, which he'd found on my LinkedIn page. "I see you worked in public relations for a number of years," he said. I said that I had. "And as a public relations executive, you are skilled at crafting a message to achieve a certain purpose, are you not?" I felt like he was implying that I could somehow turn my ability to write press releases into a Jedi mind trick that would convert not only the committee, but the entire state of Indiana, into gun safety advocates. I could see he perceived me as a threat—and not in a flattering way; I felt like I had a target on my reputation and he was taking aim.

That's when I realized that the committee members were planning to make an example of me and tear me down because of who I was. That realization triggered a full-on, fight-or-flight adrenaline rush response in me. Clearly, I wasn't going to be able to just read my remarks and go on my merry way, so I switched into a different gear.

I thought to myself, "These middle-aged white dudes are *not* going to mansplain gun safety to a mom after a shooting inside a school." If they were going to go after me, I was going to give it right back.

When one of the lawmakers bizarrely yelled at me that his wife had the right to carry a gun, I calmly pointed out that he was assuming a gun would make her safer when, in fact, a gun in the home is more likely to be used against a family member or by someone in the family to kill themselves. And that's when the committee chair said to me, "Please don't speak unless we've asked you a question," even though the NRA lobbyist talked out of turn all the time. The lawmaker went on to accuse me of not loving my country because I supported background checks on every gun sale.

As I fought back the urge to roll my eyes, I reminded myself that I wasn't just some lady who was there by accident; I knew the data, and I knew that as a mother I had an ingrained moral obligation to defend those who would potentially be harmed by this bill. From my experience as a parent, I also recognized this behavior as bullying, and I needed to do exactly what I had always counseled my kids to do if someone tried to push them around—stand up for myself. So I explained to the lawmakers just how many guns were bought and sold in unlicensed sales every year, and how many of those sales end up putting guns into the hands of criminals. I could tell that my staying calm in the face of their bizarre outbursts only made them look worse.

I wasn't the only one who thought so. During my testimony,

one of the only woman lawmakers in the room tweeted, "Bullying . . . it doesn't just happen in schools." She later told reporters, "I thought the behavior of some of our committee members and advisers was a little over the top."

I'd known that taking on the gun lobby was going to be tough, but it hadn't crystallized until that moment just how much toughness the fight was going to bring out in me, and in volunteers across the country. Even though my instincts served me well in that moment, it was clear I needed to practice my debating skills. Especially because that bill, which was opposed by the state associations of teachers, superintendents, and school boards and supported by the NRA, passed the legislature and was soon signed into law by then-governor Mike Pence.

Speak Up, Even If Your Voice Shakes

On that day after Sandy Hook when I was standing at my kitchen counter, if you had told me that what I was about to do would involve public speaking, I wouldn't have done it. But now, six years later, I've done so many media appearances and given so many speeches that it's no big deal. At all. I can speak off the cuff and feel confident doing it. Having that huge fear fall away has taught me so clearly that whatever scares you loses its power once you just start doing it.

I've also come to learn that well-timed vulnerability is just as persuasive as—if not more so than—having a snappy comeback. Sometimes you have to speak up to push back against bullies, as I did at the Indiana statehouse that day. And sometimes, speaking up is more about being vulnerable than it is about being tough. You may not feel like a badass if your voice and hands are shaking or your knees are knocking, but you absolutely are.

A perfect example of this happened in early 2018 when Moms Demand Action volunteers in Annapolis, Maryland, showed up at a rally to support a bill to take guns away from people who had been convicted of domestic violence. At first, a group of gun-rights activists who said they were there on behalf of the NRA heckled them. But then something astonishing happened. Former police officer Angela Wright shared her story about how her abusive husband used to torture her by forcing her to play Russian roulette with a loaded gun.

Standing at the mic, Angela said, "I would often wake up in the middle of the night or in the morning with the sound of 'spin, click, spin, click' as he played Russian roulette with a gun to the back of my neck." She went on to tell how one day her husband came at her with a gun with the intention not just to scare her, but to kill her. She called 911, and her former colleagues arrived in time to save her life. The NRA crowd was quiet.

Angela didn't shout, didn't throw witty jabs at anyone. She simply shared her story. Her bravery inspired the Moms Demand Action volunteers in attendance to walk up to the NRA supporters after the talk was over and open a dialogue.

Captain Tyrone Collington, commander of the nearby Takoma Park Police Department, was one of the Moms Demand Action volunteers who instigated that conversation. A military veteran, Captain Collington was shot twice before he joined the force when he tried to stop a murder in his neighborhood.

He nearly bled to death. "We walked over to introduce ourselves to the group so that we could give them a clear understanding of what we were trying to achieve so that they didn't think we were trying to take away their guns," he says. "I told them I'd been in the military for twenty-two years and have been around guns all my life; in fact I'd just purchased an off-duty weapon. This effort wasn't about restricting the Second Amendment, it was about keeping guns out of the hands of dangerous people." As a police officer, Captain Collington also shared how domestic abuse calls are the most common situations officer respond to, and also, when the domestic abuser is armed, the most likely source of officers being killed in the line of duty.

It worked. The NRA supporters changed their tune. One of the men was quoted in the local media as saying, "We support the moms in this. We are all against domestic abusers. We believe they're criminals. They shouldn't have handguns, or guns of any kind."

By the end of April 2018, the Republican governor of Maryland, Larry Hogan, had signed several bills that revamped the state's gun laws: one that banned bump stocks, which essentially turn semiautomatic rifles into automatic weapons; one that instituted a red flag law, which makes it easier to temporarily remove guns from someone who is armed and appears to be a danger to themselves or others; and one that allowed law enforcement officers to remove the guns owned by people who are under a restraining order, even before their convictions. And after he did, he publicly thanked Moms Demand Action for their support.

I know how impossible it can feel to find common ground

with people who appear to be on the opposite side of an issue from you. But at Moms Demand Action, we see these kinds of mutual understandings happen all the time. So much good can come out of conversations. The consensus that was reached in Maryland that day would never have happened if Angela had let her fear keep her quiet. Your voice may shake, but sharing your story and your point of view can move mountains.

How to Deal with Bullies

I've really had to build my tolerance for fear while doing this work—big time. And not just fear of looking like a fool, but of being physically harmed. Within twenty-four hours of starting the Facebook page, I started getting death threats and threats of sexual violence. My email was hacked; my Facebook photos were downloaded and distributed publicly; my phone number and home address were shared online; my children's social media accounts were broken into and the names of their schools shared online. Soon I started receiving creepy Son of Sam–esque messages made with letters cut out from newspapers. The underlying message: Stop talking about guns, or we'll harm you or someone you love.

I called my local police, but they weren't much help. "That's what happens when you mess with the Second Amendment," one officer told me. Thanks, guys. I did get a restraining order against the person sending the letters, at least.

As Moms Demand Action began to grow and win in statehouses and boardrooms, the threats and outrage from gun extremists grew more intense. After we joined forces in 2014 with New York City mayor Michael Bloomberg's organization Mayors Against Illegal Guns (under the new umbrella group Everytown for Gun Safety), his security team insisted I travel with a bodyguard. My security person's main job is to scout out local hospitals wherever we go so that, in the event that I do get shot, he knows exactly where to take me. Frankly, I know this work carries a risk to my safety. And I don't want to sound glib—I've met so many survivors since starting Moms Demand Action, and there's no greater horror than having a loved one killed by gun violence. But I truly believe that we're on the right side of history. At some point, you have to choose your regrets, and I would regret *not* doing this work much more than getting injured or killed in the line of duty.

I've encountered plenty of menacing trolls online, via the Moms Demand Action social media channels, too. Some of these are targeted toward me personally, but most are more generally misogynistic; every day they have posted horrific comments about women and gun violence survivors. In the beginning, I blocked each and every one of them myself. It felt futile, though. One day, I'd had enough and was lying on the floor of my closet, crying. At that moment, I got a phone call from a woman who lived nearby in Indianapolis. She could see on the Moms Demand Action Facebook page how much hate was coming our way. She said, "I'm disabled. I'm home all day. Let me block these people for you."

She performed that role for years, and I am eternally grateful for her faithful efforts to keep our Facebook page a safe space and build a team of people to help in that effort (in a perfect example of how no one has to take on everything herself). I love her for reaching out in that moment of need—it's just more proof that when you step up and stand up for what you believe in, you'll find your soul sisters.

You Don't Know Until You Try

Our volunteers have been heckled at marches, rallies, and even at the grocery store while wearing their red Moms Demand Action T-shirts. Some have reacted by using only their first or maiden names when talking in public about their gun safety activism. Some have been embarrassed or angered by the confrontations but have turned those feelings into fuel to be even more fearless. And others have been able to turn their would-be bullies into supporters—or at least neutralize them with charm.

A perfect example of this happened at a January 2016 meeting of the then-new Northern Kentucky Moms Demand Action local group. Because this was a new group, the organizers wanted to meet in a safe and inviting space to attract as many people as possible, so they chose the public library in Covington, Kentucky (just across the Ohio River from Cincinnati).

As soon as the leaders pulled into the parking lot, though, they sensed trouble. Michele Mueller, who at that time was the head of the Ohio chapter and who was there that night to help get the new group off the ground, saw a pickup truck in the parking lot with a sticker that read, "If you say guns kill people one more time, I will shoot you, and you will coincidentally die."

"I thought, 'Uh-oh,'" Michele recalls.

Sure enough, as soon as they walked in, they saw ten to

twelve men, many carrying guns, standing at the main desk, looking for our meeting. The library couldn't do anything about it, nor could the police, thanks to a 2013 Kentucky law that allowed open carry in any city-owned property. What these men were doing was perfectly legal.

And while they might have thought they could prevent the Moms Demand Action meeting from happening, they soon found out how wrong they were. Our volunteer leaders took a private moment to talk about whether they should hold the meeting or not. "We said to each other, 'The people of northern Kentucky have been waiting for a new group to start,'" Michele remembers. "So we set up our table, put out our buttons, stickers, and sign-in sheets, and opened the door." The gun extremists tried to walk right past the table, but the meeting leaders insisted—nicely—that they sign in and take a sticker.

When everyone was seated, the volunteers explained that they'd give a presentation and then there'd be time for questions. Everyone sat respectfully through the meeting—except for one mother and a grandmother who came into the room with young children, saw the guns tucked in waistbands and strapped to legs, and left.

When the presentation portion was over, several of the gun supporters came right up to the volunteer leaders and began accusing them of telling lies and asking them why Moms Demand Action thought they didn't have a right to carry guns. Michele recalls: "The gentleman who came up to me said, 'Everything you just put up on that screen are lies.' I told him, 'No sir, that was all truth. Everything we do is data driven, that's how we determine strategy." One of the men was so angry he was visibly shaking, so Michele and the Kentucky leaders pivoted. "We take photos at every meeting, so I told the man I was talking to, 'You hold this end of our sign and I'll hold the other side.' He tried to give the sign back to me, but I said, 'Hold still, it will be over in a minute.'" Another Moms Demand Action volunteer snapped a photo. Then Michele and the Kentucky leaders thanked everyone for coming and they all cleared out of the room.

Showing that kind of unflappable hospitality is such a mom thing to do. While it's nice to be friendly, it's also—you guessed it—badass to not only be cool in the face of extremism, but to figuratively disarm your opponents by being nice.

"No matter the intimidation or attempts to ridicule us and tromp our message, we stayed focused on our goal for that day—to initiate a brand-new group of Moms Demand Action volunteers willing to raise their voices—and it brought us through," Michele says. "They felt the power of their guns, but we felt the power of our sisterhood (and our placards)."

Don't Back Down; Double Down

I've learned that you basically have two choices when someone bullies, confronts, or attempts to intimidate you: either be quiet and fade away, or double down. To me, doubling down means pushing back and continuing to speak out. It certainly helps to have facts at your fingertips to debunk detractors. But mostly what you need is a large pair of metaphorical ovaries—in other words, courage.

Ironically, I learned the power of doubling down from the NRA. After all, that's exactly what that organization did after the Sandy Hook school shooting when Wayne LaPierre blamed just about everything under the sun for the slaughter of those innocent children and the educators who died trying to protect them—everything *except* for guns—and presented the NRA's dystopian vision of a United States where every good guy has a gun.

Even though I had been a public relations executive and knew the power of redirecting the narrative, I'd never seen it done to this level before. At first I was shocked and angered by the NRA's gall. But once I decoded its strategy, I started to use it.

For example, whenever there's a shooting tragedy and a resulting public outcry about changing gun laws, the NRA and lawmakers beholden to them always say it's too soon to talk about it—that it is inappropriate to "politicize" the shooting. Early on, Moms Demand Action decided we'd speak out loudly after each and every national shooting tragedy and those that make the local news because that is exactly the right time to talk about something that is senseless, preventable, and—because it's America—political. At first, our tweets calling for action right after a shooting got huge blowback from NRA pundits and lawmakers—mostly Republican—who would invariably say, "This isn't the right time."

Dana Loesch, a longtime NRA spokesperson, has called me a "ghoul" and tweeted "How dare you??" whenever I've suggested right after a shooting that gun laws need to be changed. She took it to new heights in 2015 after the San Bernardino shootings when she called Moms Demand Action volunteers "tragedy dry-humping whores" on her TV show. (I'm still not sure what that means.)

By taking a page out of the NRA's playbook, we now turn

on its head the rationale that it's inappropriate to talk about changing gun laws on the heels of a tragedy by refusing to be silent. After all, after a plane crash, you don't say, "We can't look for the black box yet, it's too soon." Right after tragedy strikes is *exactly* when you talk about preventing the next tragedy. You can be proactive as you mourn—as I've said, raw emotions are huge motivators for change. To wait until they've passed is to ignore a powerful tool. It's also a recipe for missing the moment, and that's exactly what the NRA has banked on—that by delaying talking about changing gun laws until a "more appropriate time," the news cycle will have moved on and interest will have waned.

Not anymore. You can see evidence that the culture has changed by looking at the teenage survivors of the February 2018 mass shooting at Marjory Stoneman Douglas High School in Parkland, Florida—they dealt with their grief in large part by engaging in activism right away, and they made huge strides.

I know how tempting it is to wait "until things calm down" before you take action. But if you do, you lose some of the strength and resolve that your being upset lends you. Also—especially with the vicious speed of the news cycles these days—you will be waiting a long time. Too long. When it comes to gun violence prevention, putting action off for another time means that lives will be taken.

Be Undismissable

Another bullying tactic that's been a favorite of the NRA is to try to belittle me and Moms Demand Action volunteers by portraying us as 1950s-era housewives. In fact, it ran a feature article on me in its magazine called "Not Watts She Seems," complete with an image of my head on a paper doll with an iron, a rotary phone, a feather duster, and a cast iron skillet filled with crinkle fries floating around it. (Really, does anyone use a feather duster?)

As much as I might like to think that they target us solely because we're so successful, they also do it simply because we're women. Despite the fact that the NRA appears to want to sell guns to more women—they launched NRA TV, which airs shows clearly aimed at a female audience such as *Armed & Fabulous* and *Love at First Shot*, in 2013 and hired women as spokespeople, including Dana Loesch—it's notorious for its sexist views.

In 2014, the NRA posted a video titled "Beauty Shots" comparing a woman's body to an AR-15. Worse yet, the video was released three days after a gunman at the University of California–Santa Barbara targeted sorority houses in a clear attack on women. Talk about tone deaf!

The NRA's leadership has made a whole bevy of offensive remarks about women: when then-governor Sarah Palin considered NRA board member Wayne Anthony Ross for attorney general in Alaska, reports surfaced that in 1991 he allegedly defended the rights of men to rape their wives, saying, "If a guy can't rape his wife, who's he gonna rape?" And during a debate on the Equal Rights Amendment, he apparently said, "There wouldn't be an issue with domestic violence if women would learn to keep their mouth shut."[1]

Another NRA board member, the singer Ted Nugent (of "Cat Scratch Fever" fame) called Hillary Clinton a "toxic cunt" and a "worthless bitch," Senator Dianne Feinstein a "worthless whore," and both Sarah Brady and Janet Reno a "dirty whore." He also said a feminist "is some fat pig who doesn't get it often enough."[2] I could go on. And the NRA *invited him to address the Women's Leadership Forum* at the NRA national convention in 2018.

Even more insidiously, the NRA plants seeds that women are not to be believed. They have been quick to suggest that some women will make up false claims of domestic abuse in an effort to get guns taken away from their abusers,[3] and that many campus rapes are just drunk hookups that a woman decides she regrets the next morning (a statement a guest on NRA radio host Cam Edwards's show *Cam & Company* made, to which Edwards agreed, on September 5, 2014).[4] As if any young woman would willingly open herself up to the harsh judgment our culture affords rape victims if it weren't something she had actually lived through.

Whether through intentional targeting or a more subtle doubting of our motives, plenty of people question a woman's ability to do anything important. And a big group of mostly middle-aged women? Forget it. Let's face it, we're basically invisible in this culture.

In 2018, I was interviewed for a podcast and the host repeatedly asked, "You understand why people don't believe you did this from your kitchen table, right?"—as if it were impossible to be a stay-at-home mom and create something powerful. Later in the interview, when I relayed how I had interviewed multiple potential organizations to partner with before we joined forces with Mayors Against Illegal Guns to create Everytown for Gun Safety, he was incredulous, saying,

"*You* interviewed *them*?" I replied in the moment that I didn't know whether I should be offended or flattered by that question, but really, if I had been a man, would he have asked me any such questions?

Our motto for dealing with these micro- and macro-aggressions is "Keep going," a mantra I cover in greater detail in Chapter 11. People can say or assume whatever they want, but we know who we are, and we aren't going to stop doing all the things we know to be effective just because some group of (mostly) men can't deal with the fact that we're women—and we're powerful.

We also, honestly, often get a kick out of it—especially considering that I have never been a cook and never, and I mean *never,* iron. And drinking boxed wine? If only we could take it with us to legislative sessions, we'd be able to stay there for days.

Remember, the very thing that our detractors think makes us irrelevant—that is, the fact that we are moms—is our greatest strength. Whether you handle bad behavior with humor or with directness, the best way to take the sting out of attempts to belittle you are to call them out. Don't let it lie. And don't let it quiet your voice. Speak up even louder.

Being a Badass Isn't Always Sexy

Sometimes badassery is unglamorous. Try sitting in a statehouse for seventeen hours at a stretch—that's how long Moms Demand Action volunteers sat in the Kansas statehouse in 2017 when the legislature was considering a bill to lower the age for concealed carry (it failed). Or in Oklahoma, when our volunteers sat in the statehouse to try to stop a permitless carry bill (it passed, but the governor vetoed it). It is tedious. It is uncomfortable. A million things in your regular life aren't getting done during that time. Yet showing up and staying put can be a seismic act.

Our volunteers spend their time in statehouse hearing rooms by tweeting, texting, talking, and trying to keep their toddlers from licking the floor. They do crossword puzzles and play games on their smartphones. One of our favorite ways to pass the time is to play a special form of bingo we created expressly for legislative sessions, where you get to cross out a square whenever you hear someone say specific words or phrases, such as "traitor," "Second Amendment," or "moms." It helps us pay attention and takes the focus off of how long we've been sitting there.

Some of our volunteers are prolific knitters who use the time to make gifts such as scarves and hats for gun violence survivors. Not everyone likes the fact that we knit. In fact, in Oregon, a gun extremist who regularly testifies against stronger gun bills once banged his fist on the hearing room table and shouted that our volunteers were being disrespectful for knitting while he talked about his rights.

If he thinks that's bad, he should brace himself: other women nurse their babies. One volunteer who has breastfed two kids, back to back, for the previous four years was initially told by security that she had to leave the hearing room "to do that sort of thing." She said, "No thank you. I'm good here." She swears there's not one legislator who hasn't witnessed her breastfeeding, and they finally gave up asking her to stop.

Whenever we can, we use our spare time in statehouses

to network: we pop into legislators' offices to see whether we can have a quick meet-and-greet; if they're not there, we get to know their staff members (and then look for them in the halls so we can show them that we're present and we're watching). We get to know the other groups that are there. Sometimes we recruit people to our cause—and sometimes those are people who walked into the statehouse as members of the opposition.

A badass isn't afraid of drudgery. Every single tedious little task counts and builds our collective power.

The gun lobby and every other force in America trying to mold the culture and laws to suit their special interests are banking on the fact that we're too busy and too easily distracted to put a sustained effort into resisting and creating change. Of course, you have to take care of yourself, your life, and your family, but a badass perseveres. And when she has to step back and take a break, a badass passes the baton. (Because there is always a fellow badass ready to have your back if you ask.) You don't have to do it all on your own—and you couldn't, even if you wanted to—but you are also capable of so much more than you think you are. The way you find that out is by trying.

After all, as I've said, every day in your own family you already perform superhuman feats—and, let's face it, a long list of tedious, repetitive tasks (making lunches, doing laundry, cleaning out backpacks) that go with the territory of parenting. That perseverance, no matter how painstaking a chore, that you've honed as a mom is a big piece of what makes you a badass. I like to think of the way we can tackle a heap of to-dos as *relentlessness*. And being relentless has been a crucial piece of our success at Moms Demand Action. Every time guns are discussed in a statehouse or in Congress, we're in the audience showing lawmakers that we're watching them. We're making calls. We're posting on social media. We're showing up. Our power—just like yours—is in our passion. And our passion shows up in the midst of a fight as well as in the everyday grind.

Be a Badass in Your Own Backyard

It's one thing to become more vocal online and to start showing up at your statehouse, but in my own experience and from what I hear from a lot of volunteers, it can make your knees knock even more strongly to become more visible as an advocate in your own day-to-day life.

When you take a stand for something and start devoting time and energy to making a change, it can send ripples out into your world that knock a few relationships loose. In my friend circle, in my community, and even within my family, some people supported what I was doing, and others didn't. (I'm sure you can guess which group I still have strong relationships with.) My uncle, for instance, is a gun extremist and started trolling me online to the point that I had to block him.

Sometimes those ripples can affect your kids, too. In October 2014—less than two years after Sandy Hook—my daughter came home from school and said, "You're not going to believe what happened today." And she was right. The school had sponsored a Halloween costume contest, and a kid had shown up wearing camo and carrying toy AR-15s. Emma told me that when she passed him in a stairwell, she was terrified. He looked exactly like a school shooter. Worse yet, this was a kid we suspected had made online threats to both me and my daughter and had bragged about the fact that his parents had given him guns. Incredibly, *he won first place in the costume contest.*

I called the school and emailed the principal—who was male. He said I was overreacting and that the school had sanctioned the costume. School administrators told Emma that I was being overly sensitive. But I knew that I was far from the only parent (or child) having this reaction (of course, all the other upset parents knew who to come talk to). It seemed like such a commonsense thing—that in this day and age it's not okay to have kids carry toy guns in school. I couldn't believe the school's reaction.

I posted about it on Facebook, and an Indiana Moms Demand Action volunteer reached out and said that her husband was a lawyer and he'd love to help me. He composed a letter that said if the school didn't course correct immediately, school administrators would potentially face litigation. I never heard back from the school directly about the matter, but a few days later, a letter came home in the kids' backpacks that said that toy guns were no longer allowed in the school. They even banned toy swords in the school play, which was perhaps a little over the top, but it was good to see that they had received some kind of message. Sadly, that message was probably the one from the male lawyer; without that letter, I was perceived as just another hysterical woman. Regardless, I'm glad no other kids at that school will have to experience the fear my daughter did.

I know it wasn't easy for my kids to be the ones with the mom who was stirring this particular pot, even though they believed in what I was fighting for. It's vital to remember that the most important way we teach our kids about honoring their values is by modeling those values. If we only talk the talk but don't walk the walk, we're teaching them to not take action. So press on!

Whether your bravery manifests as sharing your truth, confronting bullies, taking on your kid's school, or doggedly doing one thing and then the next thing and then the next, whenever you choose to go about creating the change that matters to you, you officially become a badass. That's why so many of our members have gotten "One Tough Mother" tattoos—once you recognize how much you're capable of, you're emboldened to be loud and proud about what you care about.

4

Losing Forward

I've always been a go-getter and a list-checker—the exemplar of a type-A woman. As evidence, by the time I was twenty-nine years old, I had three children and was a corporate vice president. If there was one thing I was pretty confident I knew, it was how to succeed. But when it came to going up against the literal big guns, I had no idea just how much relearning I would have to do.

My first lesson was delivered courtesy of the Manchin-Toomey amendment. This bipartisan bill was proposed in 2013 shortly after Sandy Hook by Senators Joe Manchin, a Democrat from West Virginia, and Pat Toomey, a Republican from Pennsylvania. It would have closed the background check loophole that allows unlicensed gun sales, mostly online and at gun shows. Passage of this bill would have been an important acknowledgment from Congress that our nation was in the throes of a crisis, and a demonstration of our nationally elected officials' commitment to acting in the aftermath of such a horrific tragedy.

It also became Moms Demand Action's first major, coordinated initiative as a newly formed organization. I'd received a phone call from a White House staff member who said the administration wanted our support as they worked to push the new bill through Congress. It was a thrill just to be recognized by the White House. It was also a relief to have such a clear goal: Pass the bill. Enact change. Achieve success. Go back to our normal lives. Simple.

Or so we thought. How could US lawmakers not act after twenty first-graders and six educators had been slaughtered in the sanctity of an elementary school? Other nations, like Australia and Scotland, had tightened laws and made it more difficult for dangerous people to get guns after shooting tragedies of their own. Surely, US lawmakers would do the bare minimum and make sure that every single gun buyer was required to undergo a background check.

We threw everything we had into championing the bill. Our volunteers began calling, emailing, and meeting with members of Congress in the districts where we lived, asking them to support Manchin-Toomey. We stood behind President Barack Obama and Vice President Joe Biden during a press conference at the White House urging Congress to act. And, in just a matter of weeks, we organized a national advocacy day called "Moms Take the Hill" in Washington, DC.

The plan was for as many volunteers as possible to travel (on their own dime) from their home states to DC to lobby their lawmakers in person. We figured out every detail—the hotels, the transportation, the scheduling with lawmakers' staff members—on what seemed like hourly conference calls, attended and run by women who had never even spoken to each other before but who were now bound together by a shared mission.

When the day arrived, just four weeks after the initial idea, I found myself at a press conference at the Capitol in the Gabriel "Gabe" Zimmerman Meeting Room, named for the man who was killed in 2011 while doing his job outside a Safeway in Tucson, Arizona, by a gunman who had attempted to assassinate then-Rep. Gabrielle Giffords. Nearly a dozen congresswomen attended to throw their support behind Manchin-Toomey and Moms Demand Action. Just before it was my turn to speak, then-Rep. Nancy Pelosi put her arm around me and whispered in my ear, "Moms will get this done."

On April 17, 2013, the day of the Manchin-Toomey vote, I was able to finagle invitations for myself and a few other volunteers to sit in the Senate gallery to watch the vote. Even though we were hopeful that the senators who were on the fence would fall our way, media were reporting that the vote could go in either direction. Even my own Indiana senator Joe Donnelly hadn't revealed publicly how he'd vote, despite the pressure we'd put on him for months.

After the senators made their remarks for and against Manchin-Toomey, I was heartened when Senator Donnelly, a Democrat in a red state, cast a "yes" vote. But I was stunned when Senator Heidi Heitkamp, a mom and a Democrat representing North Dakota, voted against it. And then other Democratic senators followed her lead, including Senators Mark Pryor, Mark Begich, and Max Baucus.

In the end, Manchin-Toomey failed by just six votes, fifty-four to forty-six. Four Republicans sided with the majority of Democrats to support the measure; five Democrats opposed it.

As Vice President Biden dejectedly announced the outcome, I heard yelling from the opposite side of the gallery where many gun violence survivors were sitting. Pat Maisch, the sixty-four-year-old gun violence survivor and badass who had taken down the shooter outside that Safeway in Tucson with her bare hands, was yelling, "Shame on you! Shame on you!"

Yelling in the Senate gallery is a big no-no. Observers are supposed to speak in hushed tones. Security swooped in and removed Pat from the gallery and sat her in a chair outside the door where they—wait for it—performed a background check to determine whether she was a criminal.

Meanwhile, in the White House's Rose Garden, President Obama and Vice President Biden held a somber press conference. "All in all," President Obama said, "this was a pretty shameful day for Washington." Gun violence survivors, including family members of the Sandy Hook school shooting victims, wept silently in the background.

Learning from Defeat

In the hours and days after the Manchin-Toomey vote, I wondered whether Moms Demand Action would survive this blow. We'd expended a huge amount of energy to pass the amendment. And we had failed. Our best hadn't been enough. Maybe the pundits were right; maybe there wasn't enough will to stop the carnage. Maybe the gun lobby was just too powerful. I seriously wondered whether we should acknowledge defeat and disband the fledgling grassroots network we'd started and galvanized.

But I couldn't stop thinking about what Senator Heitkamp told reporters after the vote. She said the reason she voted against the bill was because she'd gotten so many calls and emails from gun extremists who felt that any new gun laws were an infringement on their Second Amendment rights. Even though polls showed that more Americans supported Manchin-Toomey than opposed it, lawmakers were letting the more vocal minority shape our gun laws.

After several days of soul searching, I realized that even though Manchin-Toomey had seemed like such a no-brainer, the Congress we'd had the day before the Sandy Hook school shooting was the same Congress we had the day after—one that was beholden to the NRA and its deadly agenda. If we wanted to change the nation's culture of gun violence, we'd have to change our lawmakers one at a time. And that was going to take many years and several election cycles. In other words, this would be a marathon, not a sprint.

I also knew that although we'd lost the vote, we'd made some incredible strides. All those marches where we'd shown up wearing our Moms Demand Action T-shirts, all those phone calls and emails we'd made to lawmakers, all of the social media posts that got shared and helped build our following—they all mattered. We had learned *so much* about how to organize. We hadn't scored a touchdown, it's true. But we'd made a lot of first downs.

Yes, we're all waiting for a cathartic moment in Congress when lawmakers finally stand up to the gun lobby and close the loopholes that allow dangerously easy access to guns for abusers, criminals, and the mentally ill. And yes, bad bills get passed every month that allow unsafe laws to continue. What the Manchin-Toomey loss showed us so clearly is that Congress isn't where this work begins—it's where it ends.

After that tough loss, we turned our focus to making changes at the state level. Yes, in some ways it's a bigger task because there are fifty states and thus a lot of fronts. But it's also more doable to effect change at the state level—something we saw right away.

As the Manchin-Toomey amendment was working its way to a vote, Connecticut, Delaware, and Maryland were working to pass sweeping gun reform legislation. After the national bill lost, Moms Demand Action volunteers jumped in to play a huge role in helping those state reforms become law. They made thousands of calls to their elected officials, sent them thousands of emails, and showed up in person for hearings and meetings. After all three gun reform packages became law, the governors of each state publicly thanked Moms Demand Action for helping them pass these lifesaving bills, providing us with significant political capital—and more momentum to keep fighting.

The Upside of Losing

Thinking about all these small-yet-important gains forced me to change the way I thought about winning and losing. I had always known that changing America's gun laws wouldn't happen overnight, but up until that point I hadn't completely understood that we needed to dig in for the grueling, years-long process of incremental progress. In any kind of advocacy, you're going to encounter losses and setbacks. In order to have the mental wherewithal, we were going to have to redefine success and accept that while we couldn't always win, we could make every loss count by *losing forward*.

To the untrained eye, losing forward can look a lot like plain old losing—especially when the good gun bills we support don't pass. But there's a flip side to protecting more Americans from gun violence, and that is working to make sure bad gun bills don't get the votes to become law. And it turns out, we're really good at that.

In fact, we have a 90 percent track record of killing bad NRA-supported bills every single year—bills that would allow guns in elementary schools or on college campuses, would legalize permitless carry, or would make "stand your ground" (a questionable premise that essentially makes it okay to shoot first and ask questions later in any situation where you perceive a threat) a valid legal defense. Before Moms Demand Action came along, no one was providing any opposition to these bills; now we do that.

Of course, we do still lose. Some of those bad bills do become law. No matter what cause you're championing—and especially when your opponent has an annual budget of hundreds of millions of dollars—losing is a fact of life. What Moms Demand Action volunteers have learned, though, is that losing can help chapters grow, result in stronger relationships with lawmakers, and even change hearts and minds. When we inevitably lose, we feel the disappointment, but not for long; we're too busy focusing on the gains our efforts created, even if they didn't have the exact results we were aiming for. Besides, if we gave up when we lost those battles, we'd be letting the NRA and gun lobbies win the war, and we've come too far to let that happen.

As the husband of one of our Virginia volunteers once said

to me, "Sometimes entire football games are won by field goals—not everything has to be a touchdown."

Be Dedicated to the Ends, but Flexible About the Means

In any advocacy work, doors will slam in your face—literally and figuratively. When you commit to losing forward, you know the door is just one way to get inside—you can always go through windows, too, as I said in Chapter 1. The power of losing forward comes from simply refusing to give up. While you still need goals you're working toward—ours are demanding action from state and federal legislators, companies, and educational institutions to establish commonsense gun reforms—losing forward helps you stay flexible about how to reach those goals. Challenge yourself to find a new way in any time you encounter a closed door. And really, does it matter what your path looks like as long as it gets you where you intend to go?

Flexibility in Action

The campaign by the Texas Moms Demand Action chapter to simplify "no guns" signage in Texas is a great demonstration of the power of the willingness to be flexible about your means.

Shortly after the Manchin-Toomey vote, one of our Texas volunteers, Norri Leder, stumbled upon a website called texas3006.com. This site, which still exists as I write this, lists thousands of Texas businesses and nonprofits that prohibit guns on their premises. Norri noticed a column that she didn't understand at first; it had the header "Valid" and listed either "yes" or "no" for each establishment. She soon figured out that the "yes" or "no" indicated whether the signage the business posted complied with state law. The information was listed because gun extremists were using it to determine whether or not they could legally flout company policy. If a business's signage didn't follow the letter of the law, these website users would carry their guns into the business anyway. When Norri saw the comments from folks saying things like "I carry into the Chuck E. Cheese on so-and-so street because the sign is up in English and not in Spanish," she got that hit of outrage that told her she had to try to change it.

Norri began to dig and found that in the mid-1990s, shortly after concealed carry was legalized by the Texas legislature, a rash of businesses began posting "no guns" signs. In response, the NRA and Texas State Rifle Association used their influence to get a law passed through the state legislature that they dubbed the "big ugly sign law." This law required businesses to post large signs that contained extensive boilerplate language in a specific font size in both English and Spanish. Because it was a nonstandard size, the mandated sign required professional printing, making it an inconvenient and expensive eyesore.

The best part for the gun lobby? Most businesses knew nothing about the sign law, especially by the time Norri found out about it, many years after its passage. As this website proved, hundreds of businesses were posting signs they thought were sufficient to prohibit guns on their premises when technically, they weren't.

Norri began calling every business with a "no" in the "Valid" column to ask whether their signage was compliant. She and the Texas Moms Demand Action chapter also persuaded multiple business owners to send letters to lawmakers asking for simplified signage. Using social media and direct outreach, they eventually persuaded influential state groups such as the Texas Association of Business and the Texas Restaurant Association to support legislation to change the signage requirements.

Business-friendly and gun-sense-friendly lawmakers put a bill forward to change the signage requirements, and Norri was asked to testify. During her testimony, she showed just how big and unsightly the mandated signs were by pulling out one of them, and she gave lawmakers a list of businesses in their districts that had been targeted on texas3006.com.

Unfortunately, the simplified signage bill didn't pass. We could have chalked that up as another defeat, but our dedication to losing forward motivated us to pivot our attention to the businesses themselves. Volunteers called and visited businesses all over the state to let them know about the signage requirements and to offer them professionally designed and downloadable versions of the sign, so that all a business owner had to do was email it to a printer and then hang up the finished product.

Texas volunteers also created an online tool to track businesses that ultimately put up proper signage to prohibit the concealed and/or open carry of firearms in their establishments. In the end, they persuaded more than four hundred businesses to post the ugly (but legal) signs—including grocery stores, restaurant chains, medical facilities, day-care centers, and houses of worship.

Simplified signage bills were advanced again in the 2017 legislative session. Again they didn't pass, but the issue is still very much alive in the legislature and serves as a powerful reminder that while Texas Republican lawmakers fear the gun lobby, they also fear the business community. And thanks to Norri's efforts, many, many more businesses listed on the texas3006.com website have a "yes" in the "Valid" column. Even though the legislation hasn't yet passed, we still achieved our end goal—more Texas businesses legally prohibiting open carry.

The Gift of Trying Even When There's No Hope of Winning

Some battles you just know you're not going to win. It's important to fight them anyway, because when you lose forward, you make gains that help make you stronger for the next fight. That's what happened in 2017, when a small team of Arkansas moms decided to spearhead a Moms Demand Action campaign to defeat a "guns-on-campus" bill. Arkansas is a red state, so the team suspected that no matter how effective they were, their efforts were likely to result in a loss—after all, the NRA was throwing everything it had at allowing guns on college campuses. But they were determined to be a part of the resistance; doing nothing simply wasn't an option.

The Arkansas chapter leaders decided to go down swinging. What made the road ahead even more challenging was that the Arkansas chapter had struggled to grow its membership. At that time, it was one of the smallest in the nation, with only twenty-eight volunteers. They couldn't send mass emails or texts to rally the troops, because there weren't really any troops to speak of. But although those twenty-eight volunteers weren't exactly a groundswell, they were a good start.

Eve Jorgensen, a Little Rock mother of two and full-time engineer, started posting frequently on social media to spread the word about the guns-on-campus bill and its dangerous consequences. She also worked with a small group of other Arkansas volunteers to develop calling scripts. Then she created new social media campaigns to urge Arkansans to get involved and call their legislators.

"I was shocked by the power of social media," Eve said. "I'd put up a message about a last-minute hearing on the bill, and people would see it and actually show up at the statehouse."

The press noticed, too. Austin Bailey, chapter leader at the time, gave multiple interviews on air and in print to get the word out about the fight happening at the Capitol. One article referred to the Arkansas Moms Demand Action volunteers as "scarlet-shirted mothers" who were "permanent fixtures" at the statehouse. That media coverage helped draw new volunteers to the chapter. When bigger red-shirted crowds started turning out at events, the media reported on the increase, which fueled more people to join.

This is one reason why it's important to wage a battle you know you can't win: to increase your organization's visibility and public support. Simply by showing up and refusing to give up, Moms Demand Action volunteers showed Arkansans that it was possible to galvanize around a polarizing issue in a red state.

After the bill to allow guns on campus passed the legislature, but before it was signed into law by Governor Asa Hutchinson,

Moms Demand Action volunteers met with him two times in hopes of persuading him to veto it. Each time, they brought along key campus stakeholders, members of law enforcement, and constituents to share their reasons why guns on campus was a dangerous idea. Unfortunately, these meetings didn't have their intended result, either. Days later, with the NRA's chief lobbyist at his side, Governor Hutchinson announced he would sign the bill that allowed college students and faculty to carry hidden loaded guns to class, into bars, and inside campus stadiums and dorms.

Although the unpopular bill passed, playing defense against it meant that the legislature was too occupied by the fight to move forward on other bad gun bills—a true victory in a red state. And the Arkansas Moms Demand Action chapter got a much-needed kick start—as of this writing, they have nearly two thousand volunteers and growing—and expanded to include local groups in three new cities. They also just helped elect two Arkansas Moms Demand Action volunteers to the statehouse—one of whom unseated the lawmaker who was responsible for pushing the guns-on-campus bill. We may have lost that battle, but we ended up a lot better equipped to fight the next one.

If you think there's no chance in hell you can win something, you might be tempted to not even try. After all, why pick a fight when you're pretty sure you're going to get beat up? But the biggest loss is not trying at all, because it prevents you from picking up crucial gains along the way that help create momentum—whether that's new members, more visibility, an opening to change a conversation, or a social media moment that puts you on the map. Sometimes the biggest successes come not from outright triumph, but from refusing to go down quietly.

Use Your Losses to Fuel Your Motivation

Some losses feel more personal than others. Although these can be some of the most devastating challenges to overcome, they can also provide the fodder you need to keep going. In these instances, losing forward helps you make sense of your loss and channel those emotions into action. I felt hopeless after Manchin-Toomey failed. It seemed like an overwhelming, devastating defeat, but that was also the moment that I knew we needed to become so big and so powerful that lawmakers like Senator Heitkamp couldn't ignore us again. The loss was a catalyst for our growth.

But as much of a gut punch losing Manchin-Toomey was, the sadness I felt was nothing compared to Lucy McBath's.

Lucy's seventeen-year-old son, Jordan Davis, who was black, was shot and killed in 2012 outside a gas station in Jacksonville, Florida, by a white man who was angry that Jordan and his friends were playing loud music in their car.

Although he was later found guilty, Jordan's killer claimed Florida's stand-your-ground law as self-defense. This NRA-supported bill was signed into law by Governor Jeb Bush in 2005, making it the nation's first law of its kind. Stand your ground emboldens people to shoot to kill, even when they can safely resolve a conflict by other means. And data have shown that such laws result in a significant increase in gun homicides, particularly among African Americans.[1] (The law was also used to defend the Florida man who shot and killed seventeen-year-old Trayvon Martin in 2012. As the black teen headed home from a convenience store, he was pursued, shot, and killed by a man who claimed he felt threatened. Because of stand your ground, Trayvon's killer was exonerated, sparking national outrage.)

Lucy was living in Georgia when her son was killed. (Jordan was visiting his dad in Florida while Lucy received treatment for breast cancer in Atlanta.) Despite her unbearable sadness, Lucy—whose father had been president of the Illinois chapter of the NAACP—instinctively turned her grief into activism for stronger gun laws. She became a national spokeswoman for Moms Demand Action in 2013. As part of her work, she has met with lawmakers, has testified before the Senate Judiciary Committee, has spoken at the White House, and was one of seven members of Mothers of the Movement who spoke in support of Hillary Clinton at the Democratic National Convention in 2016.

Lucy's voice as a mother of a child killed by gun violence and as a woman of color has been instrumental in advancing the gun violence prevention movement. While I was writing this book, Lucy ran for Congress in her home state of Georgia, and to my everlasting amazement and delight, she won and will be joining the 116th Congress in the US House of Representatives. It was her first time ever running for any office, and she was elected almost six years to the day that Jordan was murdered. Out of loss that no mother can bear to even contemplate, Lucy has helped move the gun safety debate forward in huge strides.

I hope you never have to bear a similar loss, but no matter

what makes you feel pain and outrage, I hope Lucy's story proves that what doesn't kill you absolutely will make you stronger.

Use Your Losses as Fuel for the Next Battle

There's another way to lose forward: even though you know you'll lose, force lawmakers to publicly debate and vote on a bad bill, and then use their own words and actions to hold them accountable in the future.

This is what Moms Demand Action did in Florida in 2016 when a bill was introduced to expand stand your ground. This bill was in direct response to a July 2015 Supreme Court of Florida ruling that said defendants must prove that they believed they were in imminent danger in order to legitimately use deadly force in response. The gun lobby wanted to lessen the burden of proof on those shooters. So they pressured lawmakers to introduce a bill that would shift that burden of proof from the defendant to the prosecutor—meaning anyone who claimed stand your ground as a defense would no longer have to prove that they felt their lives were in danger. Moms Demand Action volunteers feared that this would only further embolden people to shoot first and ask questions later, which meant more people—more black and brown people, in particular—would be shot and killed.

Led by Chryl Anderson, a longtime friend of Lucy's, a woman of color, and the grandmother of six, the Florida Moms Demand Action chapter activated a network of Tallahassee volunteers to show up at the statehouse at a moment's notice while the proposed revision to stand your ground was being debated.

Chryl learned that a Republican lawmaker reported he wasn't getting any calls complaining "about bad gun bills." She used this as motivation to make some noise and launched a campaign that drove more than one thousand calls and eight thousand emails to Florida lawmakers urging them to vote "no" on the stand-your-ground expansion.

In October, while the bill was still in committee, Lucy testified against it, invoking the memory of her son. She begged the committee not to make it harder to convict armed vigilantes like the one who had murdered Jordan. The bill was tabled.

But even winning in a politically conservative state like Florida doesn't always feel like victory. "I knew the bill would be back," Chryl said. And she was right; it came back in 2017. By then, a Tallahassee volunteer named Michelle Gajda had taken over leadership of the Florida Moms Demand Action chapter, and she knew she was stepping into a losing battle.

"There was no way in hell we were going to beat back the stand-your-ground expansion this time," Michelle said, "but I was determined that we would lose forward by holding our legislators publicly accountable for passing a bad bill."

Michelle continued to cultivate the rapid-response team Chryl had built. She made sure that Moms Demand Action volunteers were ready at a moment's notice to throw on their red shirts and sit in the statehouse, waiting to testify; have conversations in hallways; or simply make eye contact with wavering lawmakers. "We provided cover for lawmakers," Michelle said. "They could point to our volunteers in the crowd and say, 'Look, these constituents are extremely opposed to this bill.'"

Another way Florida volunteers lost forward that year was that they evolved from being community activists to becoming policy experts. "Legislators learned they could come to Moms Demand Action for reliable research and information," Michelle said. "They trusted us to help them make the argument for gun safety."

When the stand-your-ground expansion bill went to the floor of the Florida House for a final vote, dozens of Moms Demand Action volunteers showed up to bear witness to the bad behavior of lawmakers who clearly cared more about their NRA ratings than their constituents' safety.

"Many lawmakers of color testified about the deadly combination of racism and stand your ground," Michelle said. "It was devastating to hear representatives say, 'If this passes, people who look like me are going to die.'"

When the bill was voted through, many of the Florida Moms Demand Action volunteers stepped out into the hallway to cry. They knew they'd lost an important battle, one that would have a death toll associated with it. But they also knew that suffering losses is an inevitable part of any war—at least now they'd gotten one out of the way. Each year, there are other stand-your-ground battles to fight. Having lost this particular battle in Florida will help secure other future wins, because every loss gets you closer to a victory. Whatever fight you are taking on, the irony is that in order to progress you need to lose, and to be willing to do it again and again. It's the only way to win.

5

Use Your Bullhorn

When I went searching for the gun safety version of Mothers Against Drunk Driving after Sandy Hook in 2012, the only thing that remotely resembled what I was looking for was the Million Mom March in 2000. While that protest drew an estimated 750,000 people and spawned two hundred local chapters in the months after, it eventually fizzled—in part because NRA-friendly George W. Bush was elected president and the gun lobby began a meteoric rise in influence in both lawmaking circles and the public mind. Timing can play a huge role in whether your moment turns into a movement. And a crucial piece of what made the time ripe for Moms Demand Action to take off was something that neither MADD nor the Million Mom March had access to when they were founded: social media.

Without this technology, Moms Demand Action probably would never have begun—after all, we started off as a Facebook page. And we certainly never would have grown to more than five million supporters, a chapter in every state and DC, over seven hundred local groups, and thousands of volunteers taking leadership roles in their community without it. Our successes prove the words of World Pulse CEO Jensine Larsen: "A woman with a laptop can be more powerful than a man with a gun." Social media have not only helped me find my people, they have helped me find my voice. And they have helped me see that the traditionally feminine talents of communicating, networking, listening, and befriending are formidable forces of change.

Although anyone of any gender, race, class, or country can use social media, I view social media as a woman's bullhorn. Because on social media, there are no gatekeepers. And the gatekeepers in mainstream media—including television, print, and online news—tend to be male, and not that interested in covering the feats of women. I feel this struggle every day as other activists—often male ones—who are fighting courageously for just causes but who are not necessarily more experienced or effective than I am are invited to be interviewed on news magazines, late-night TV talk shows, and popular political podcasts.

Even when I do get interviewed, I'm often held to a different standard. I was recently interviewed by a major daily newspaper for nearly an hour and explained all of the wins Moms Demand Action has had as an organization. The male reporter chose to quote only my male colleague, saying he'd been "more concise" during their interview. When I was interviewed on a morning national radio show about school shootings along with two men, Twitter users pointed out that the host—a man—had asked me only about the emotional aspects of such shootings; he asked the men about policy and politics.

But thanks to social media, I can reach a large audience and bypass (mostly male) gatekeepers: all told, with my personal Twitter and Facebook accounts plus all the various Moms Demand Action accounts, my reach online is just about four million people, which is the same number of people on average who watched *The Late Show with Stephen Colbert* each night in the first quarter of 2018.[1] (I'd love to be on your show, Stephen!) We don't need to wait for men to let us tell our stories anymore. We have to pull whatever limited levers of power are available to women, and social media are big levers.

I know that some of you are reading this and thinking, *Really? Do I have to?* I understand that social media are not for everyone, and that using social media has its perils: it can be a time suck; it's easy to become tethered to your phone or laptop as you check for updates, such that your work seems to invade all parts of your life; and there's also the very real danger of being attacked for sharing your views or getting into heated and upsetting exchanges with people you may or may not know.

I definitely believe that the benefits of social media far outweigh the potential harm, but if you just aren't a social media person, you don't *have* to use them. But you *do* have to decide how you'll get and share information. (Texting is a great way to do that, one that Moms Demand Action volunteers use to alert other volunteers about events and to stay in contact with each other.) And you'll have to find other ways to contribute, either with your time—showing up in person to meetings and events—or your money. Maybe by the end of this chapter, though, when you've heard all the incredible things social media enable and empower us to do, you'll be swayed to give these tools another try.

Use Social Media to Give You Liftoff

If you ascribe to the belief that you should build the plane as you fly it, social media are the wind beneath your wings. They can put you on the map even before you exist in the real world; that's certainly how that first Facebook page worked for Moms Demand Action.

Facebook was our first home, and for good reason: it's where so many moms hang out. We go there to share stories, family photos, interesting articles, and funny memes—it's how we stay connected to friends and get to know new people better. And that's exactly why Facebook was so powerful for Moms Demand Action—women were already using the platform and felt comfortable there. In our early days, our volunteers started sharing events and interacting with each other on their personal pages as well as on Moms Demand Action's public Facebook group page. It helped us become friends as well as fellow activists.

And while Facebook is still the largest social media platform, it may or may not be the most natural fit for you—or you may have some reservations about Facebook and privacy issues. Before you proclaim an online home, ask yourself, Where do most of the people who will be receptive to your message hang out? That's where you'll want to be. That said, Moms Demand Action definitely had some learning to do about how to use Facebook for activism and not just social connection, specifically around what to make public versus what to keep private.

Obviously, we have a public Facebook page where we share news articles about and statistics on gun violence, promote our initiatives (such as our recent Gun Sense Candidate tool that is our answer to the NRA's rating system for political candidates), and encourage people who find our page to become active volunteers. But we learned pretty quickly that in addition to our public pages, we needed private Facebook pages, too, because gun extremists used our public posts to learn about upcoming events and then crash them—often while carrying long guns, as happened in Kentucky. Now, each state has its own private Facebook group, as do some of the larger cities. There are also private groups for Moms Demand Action leadership from all over the country to exchange ideas, share progress, and get support.

In fact, Facebook is how I got connected to Sarah Brady, founder of the Brady Campaign to Prevent Gun Violence and wife of Jim Brady, the White House press secretary who was shot and disabled during the assassination attempt of Ronald Reagan in 1981. Before she died in 2015, Sarah was an icon of the gun safety movement. I was thrilled when she friended me on Facebook, and pleasantly shocked when shortly thereafter she sent me a message saying, "You need to get a Twitter handle for your organization right away." That's how un-savvy I was about Twitter at the time—I needed someone from an older generation to encourage me to start using it.

I'm so thankful for Sarah's advice (and hope you'll follow us: @MomsDemand). It was the right move because Twitter is more political than Facebook: politicians use Twitter to connect with their constituents in a quick and public way, and Americans use Twitter to keep up with politics. A 2018 study by the Pew Research Center showed that 14 percent of the public say they have changed their views about a political or social issue in the past year because of something they saw on social media.[2] That can include the politicians themselves!

Twitter and Facebook have remained the primary social media platforms for Moms Demand Action. But because we want to attract younger volunteers, too (and we're proud of our partnership with Students Demand Action, which is led by high school and college students and grew from a pilot program into a full-fledged movement in the wake of the Parkland shooting), we've moved onto Instagram.

Use Social Media to Be Bolder Than You Might Feel in Real Life

Despite having basically no Twitter experience when I started Moms Demand Action, I've grown to love it. It's emboldened me to be fearless and aggressive in ways that I rarely can manage in person. Twitter not only lets you share your opinions with a crowd even when you're alone in your home (the dream for an introvert like me), it helps you hone your message and develop your persona in a way that would take months or even years of the more traditional method of writing press releases and waiting to be interviewed by media.

When I first started using Twitter for Moms Demand Action, I kept my tweets pretty straightforward—sharing articles, thanking supportive lawmakers, and regurgitating our talking points. But during the run-up to the presidential election of 2016, I couldn't help myself from getting more political.

And it paid off: when I started using my own voice, I went from about twenty-five thousand followers to more than one hundred thousand in just a few months.

I've found that taking a stand for something is rewarded by Twitter users: whenever I write a popular tweet, I get spikes in followers. For example, on December 29, 2015, I tweeted at then-presidential-candidate Trump's spokeswoman, Katrina Pierson. She'd just been interviewed on CNN while wearing a necklace made of bullets, so I tweeted, "Surely Katrina Pierson [and I tagged her] wore a bullet necklace to recognize that 90 Americans die from gun violence every day." By the time I woke up the next morning, she'd tweeted back, "Maybe I'll wear a fetus next time & bring awareness to 50 million aborted people that will never get to be on Twitter." I didn't want or need to respond, but regardless, overnight I had several hundred new followers.

More recently, in the wake of the Parkland shooting, I got into a heated Twitter discussion with NRA spokesperson Dana Loesch, which started when she erroneously implied in a tweet that I wanted to ban all guns. She tweeted: "Moms Demand defines all guns as 'assault weapons' and wants to ban 'assault weapons.' You do the math," and she attached a screenshot of an exchange we'd had five years earlier. She didn't tag me or Moms Demand Action (the social media equivalent of talking behind someone's back), but friends alerted me to her tweet.

I tweeted back that she'd taken my tweet out of context—I'd been referring to how many rounds a gunman had just fired in a recent mass shooting. We then got into an hours-long back-and-forth where she kept asking me whether I wanted to ban all guns and I was asking her about the NRA's close relationship with Russia. I'd learned a lot on Twitter in five years, and one lesson (something Dana is a master at) is that I don't have to answer what someone asks me; I can create my own conversation about what I want to discuss.

In the end, I had some tweets I'm still proud of ("Does the NRA take money from foreign entities like Russia? Answer like you're explaining *Quantum Leap* to your grandma. We need to understand."), and she ended up deleting several of her tweets.

During our back-and-forth, people were tweeting me on, saying things like "Hey all, Shannon Watts is owning Dana Loesch in a Friday night tweet-out. It's better than a Friday night at the movies!"

Our responses to one another were re-tweeted by thousands, and the discussion garnered a lot of attention. Over the course of the exchange, and thanks to a concerted campaign by Hillary Clinton staffer Adam Parkhomenko, who was encouraging his followers to follow me, I gained more than fifty thousand followers.

Use Social Media to Encourage Corporate Responsibility

One of the biggest companies we used social media to influence was, ironically, Facebook—the biggest platform of them all, as well as our birthplace. From the beginning, we'd had a good relationship with Facebook. In fact, when we first started garnering attention in early 2013, the sister of Facebook COO Sheryl Sandberg, Michelle, reached out to me to organize a meeting in Silicon Valley that Sheryl attended. She was very friendly and supportive, and I was grateful to have her wise counsel.

But just a few months into doing this work, we realized that Facebook had become one of the largest online markets for unlicensed gun sales, and the platform often played a major role in getting guns into the hands of abusers, minors, and criminals.

In January 2014, we couldn't ignore the data anymore—we knew that if Facebook (and Instagram, which Facebook owned) changed its gun policies, lives would be saved. After all, Craigslist, eBay, and Google+ had already banned unlicensed gun sales on their platforms; we weren't asking Facebook to do anything unprecedented. Without any restrictions on gun sales, Facebook and Instagram were essentially hosting online gun shows every day, twenty-four hours a day.

Jenn Hoppe, our lead for corporate campaigns, had the ingenious idea of doing our own version of the ten-year look-back video that Facebook had recently released to celebrate the company's first decade. Our version showed how specific shooters had gotten their guns online, including the perpetrator of a horrific 2012 shooting at a hair salon and spa in Wisconsin who had gotten his gun through Facebook. That information went viral, garnering attention worldwide.

We also asked our followers to post comments on the public pages of Facebook and Instagram and to tweet and post on their own feeds using the hashtag #EndFacebookGunShows. They also shared graphics showing the alarming statistics of how many unlicensed gun sales take place in the United States (a survey of gun retailers by the City of New York found that private gun sellers suspect that 62 percent of the people buying online wouldn't be able to pass a background check) and the strong tie between criminal activity and unlicensed gun sales (another survey of incarcerated firearm offenders found that nearly 40 percent of them were legally barred by either state or federal laws from owning a gun, and 96 percent of that group had procured their guns from a supplier that was not required to conduct a background check).[3] We were weaponizing social media in order to get illegal weapons off of social media.

Within a month of launching our Facebook campaign, we began having formal talks with Facebook executives about the implications of the company's policies. During that time, a fifteen-year-old boy took a gun he had bought through Facebook to his high school homecoming game; and a felon was arrested in Storm Lake, Iowa, for possessing a gun he had bought through Facebook. We publicized these two incidents, which helped draw attention to the very real threat Facebook's gun policies posed.

On March 5, 2014, Facebook announced nine new policies around gun sales: most important, the company agreed to delete reported posts offering guns for sale without a background check and to no longer show gun ads to people younger than eighteen. It also agreed to force people who search hashtags such as #Guns4Sale on Instagram to acknowledge their state laws before seeing search results. It was an important start, but the policies didn't go far enough.

We kept the pressure on by providing Facebook with a report showing how domestic abusers were getting access to guns through its platform. Finally, in February 2016, Facebook announced that it would no longer allow any unlicensed gun sales on its platforms, including Instagram. In May 2017, after a murder in Cleveland, Ohio, had been broadcast live on Facebook the month before, the company hired three thousand additional employees to review and remove any content that conflicted with company policy—including advertisements for unlicensed gun sales. Unfortunately, when it comes to stamping out unlicensed gun sales via Facebook, it isn't "set it and forget it." It requires constant vigilance, and Moms Demand Action is proud of our role in keeping Facebook on its toes to prevent needless gun deaths and to review and refine its policies.

Use Social Media to Turn Cultural Conversations into Policy Progress

We've also combined our purchasing power with our social media reach to influence national retailers and restaurants to either enforce existing gun policies or create new ones. Although individual companies were our targets, we were going after something much broader—a cultural shift that brought attention to open carry. And we did it by building on the success we'd had with influencing Starbucks to ban guns in its stores.

In 2014 we launched a number of campaigns to get other large restaurants and retailers to follow suit. We never would have been able to be as effective, or as quick, in this effort without social media.

After Starbucks changed its policy, gun extremists, particularly in Texas, went on the offensive: they organized meet-ups in several different businesses where dozens to hundreds of people would show up, armed to the teeth with semiautomatic rifles, in a flaunting display and exploitation of open carry laws.

The first such event happened at a Jack in the Box in the Dallas–Ft. Worth area when members of a gun extremist group called Open Carry Texas walked into the restaurant carrying long guns. The employees were so scared that they locked themselves inside a walk-in freezer. We issued a press release, launched an online petition (we helped spread the word about the petition through Facebook ads that featured a shareable graphic—because on social media images garner even more attention than words), and tweeted photos, with the hashtag #JackedUp, of our members eating at other fast-food restaurants that had safer gun policies. Within days the company announced it would begin enforcing its policy of no guns inside its restaurants.

After that, there were similar incidents at Chipotle, Chili's, and Sonic Drive-In. And each time we led with the trifecta of a press release, a petition, and a catchy hashtag. For Chipotle it was #BurritosNotBullets; for Chili's, #RibsNotRifles; and for Sonic, #ShakesNotShotguns—and we asked our members to attach the hashtag to pics of themselves eating at competitor restaurants that had better gun-sense policies in place. During each campaign, we got thousands of signatures each day, hundreds of photos popped up all over the companies' social media feeds, and the media wrote stories about the campaign. It generally took just a couple of days for the restaurants to change their policies. (That's the other thing about social media; it can make things move very quickly.)

Of course, we wanted to counter the efforts of gun extremists. But these were more than tactical moves in the gun safety marathon. Allowing anyone to open carry a gun inside a restaurant or a store where we take our kids puts the onus on customers to figure out who the good guy is. Forty-five states allow the open carry of loaded, semiautomatic rifles in public—and most of those states don't require a permit or gun safety lock to do so. Combined with the estimate that 22 percent of all gun sales take place without a background check in the United States, this means that people in most states can legally carry rifles openly in public without ever having passed a criminal background check.

When you see someone in public with a gun, it's impossible to know whether that person is a good guy or a bad guy. Businesses everywhere have a duty to protect their employees and patrons, but even more so in states where no background checks or training is required to buy and carry guns in public. Businesses also have a duty to listen to the very legitimate concerns about the safety of their customers. Remember, American women make 80 percent of the spending decisions for their families. Businesses ignore us at their peril.

A modern proverb says, "If you don't have a seat at the table, you're probably on the menu." Social media help ensure that women have a seat at the table, particularly at tables within corporate boardrooms (remember that less than 1 percent of Fortune 1000 CEOs are women).

We knew we were affecting major cultural change when Panera Bread Company executives came to us proactively in 2014 to help them modify the company's gun policies—soon after, they announced that guns were no longer allowed inside the eighteen hundred Panera restaurants in forty states.

A company that took longer to change its policies—although it ultimately did, which is the most important thing—was Target. Gun extremists staged several open carry demonstrations inside Target stores in Texas, Alabama, Ohio, North Carolina, Washington, Wisconsin, and Virginia, yet Target's management was silent—this despite the fact that the company acknowledges that 80 to 90 percent of its customers are female and 38 percent of its shoppers have children, a share the company boasts is higher than any other discount store.

Moms have a deep bond with Target. It's where so many of us buy school supplies, groceries, trash bags, towels, and clothes for our kids, and may throw a cute pair of pajamas or some yoga pants in the cart for ourselves. It's our haven—except for when hundreds of angry gun-rights activists trying to prove a point are milling around the parking lot with semiautomatic rifles.

In early June 2014, Moms Demand Action launched a social media campaign using the hashtag #OffTarget, and we encouraged women to shop at other retailers and use the hashtag when posting pictures of themselves and/or the things they bought at Target's competitors. We also designed a printable sign that read "Don't make my family a target," which hundreds of moms tweeted along with photos of their families.

Our petition asking Target to change its policies garnered more than twenty-five thousand signatures in fewer than forty-eight hours. Still the company said nothing. Even after a loaded gun was found in the toy aisle of a Target in Myrtle

Beach, South Carolina—employees found it and reported it to authorities, but there easily could have been a much more tragic outcome—Target stayed quiet.

A month later, after our sustained campaign encouraged nearly four hundred thousand Americans to sign our petition asking Target to ban open carry in its stores, the company announced: "Starting today we will also respectfully request that guests not bring firearms to Target—even in communities where it is permitted by law. . . . This is a complicated issue, but it boils down to a simple belief: Bringing firearms to Target creates an environment that is at odds with the family-friendly shopping and work experience we strive to create."

Yes! In addition to being relieved that American families will be safer now when shopping at Target, I think I speak for pretty much each of our hundreds of thousands of volunteers when I say that I'm thrilled that I get to keep shopping there! It's hard to imagine mom life without Target.

Use Social Media to Draw Attention to the Good Guys

It's important to note that social media is also a great venue for publicly thanking companies that take important steps to promote gun safety. That's exactly what we did when, shortly after the Parkland shooting in early 2018, Dick's Sporting Goods announced it would stop selling assault weapons and high-capacity magazines and raise the age limit for buying guns in its stores to twenty-one. Then, in September 2018, Levi Strauss CEO Chip Bergh announced that the company would establish the Safer Tomorrow Fund with a $1 million gift and use those funds to support nonprofits and youth activists who are working to reduce gun violence. The company also committed to doubling the donations its employees made to organizations affiliated with the Safer Tomorrow Fund and became a founding member of Everytown Business Leaders for Gun Safety in order to encourage other senior executives to support commonsense gun reform. Moms Demand Action thanked Levi Strauss and Chip Bergh, too.

We've also had multiple occasions to thank Walmart—after the company stopped selling assault weapons and high-capacity magazines in its stores in 2015, and again when it raised the age limit for buying a gun to twenty-one in 2018. Walmart still allows open carry inside its stores, but give us some time. The company has been a force for gun sense in many other ways. In 2008 it was the first large gun retailer to adopt a ten-point code of conduct in partnership with Mayors Against Illegal Guns (another partner under the Everytown for Gun Safety umbrella). And its background check policies have been more stringent than even those of the federal government. Since 2002, Walmart has had an official policy of not selling a gun to someone who doesn't have a fully cleared national background check, even though federal law allows gun dealers to sell a gun to someone whose background check hasn't cleared after three days. This is what's technically known as a "default proceed," but more commonly known as the "Charleston loophole." That's because the mass shooter who killed nine black parishioners at a church in Charleston, South Carolina, in 2015 was able to purchase his gun even though the system had not issued a determination on his background check by the end of the federally mandated waiting period of three business days.

Not all of our social media campaigns have been what is traditionally considered "successful." We've also targeted Kroger's, Home Depot, and Staples, and none of them has, at the time of this writing, changed its gun policies. But it is still important that we try, because we can lose forward. Our campaigns alert companies' customers—many of whom care about gun safety—about store policies, as well as about state laws that allow stores to have these policies. They also give us an opportunity to tell people where they can shop instead, and let us send a little love to companies with gun sense.

Use Social Media to Speak Truth to Power

Using social media is certainly not all about picking fights. We also use them to amplify others' voices and draw attention to issues we care about that aren't getting the media attention we feel they deserve. Gun violence rarely makes the national news unless a white person is killed; we use social media with our audience to do our part to raise awareness of *all* the ways gun violence impacts marginalized communities.

We share stories about gun violence in cities and share the posts and tweets of people of color whose voices aren't being captured by mainstream media. Any time a police officer shoots an unarmed person of color, we go online and share the data that show you can't talk about gun violence without also talking about the systemic racism that can lead to it.

We also share stories about gun violence against trans and gender-nonconforming communities, which mainly affects black trans women. There's no national tracking of gun violence against black trans women, so Everytown for Gun Safety bases its data on news reports and accurate reporting of the gender identity of victims.

Another example of using social media to bring about change is how we have strategically gone after the empty sentiments of politicians offering "thoughts and prayers." At some point in 2015, after the mass shooting in San Bernardino, I just couldn't bear to hear those words one more time—and I wasn't alone. Moms Demand Action volunteers and so much of the American public were right there with me. "Thoughts and prayers" are often a direct result of the NRA's insistence that it isn't appropriate to talk about changing gun laws in the days right after a national shooting tragedy, so instead of discussing solutions, politicians on both sides of the debate are left with merely sending their thoughts and prayers.

I started using my Twitter feed to point out how gun laws had contributed to each particular shooting—maybe the shooter was a known danger to himself and others, or maybe he had a history of domestic violence, but law enforcement couldn't take away his guns because there wasn't a red flag law on the books in that state. And I'd point out how the NRA was complicit in the shooting via its efforts to weaken specific gun laws. And our volunteers and supporters used their social media platforms to amplify how each individual lawmaker was offering thoughts and prayers without the promise of change, which wasn't nearly enough to stop the next shooting. Sharing facts with a few million people is a great way to stymie hypocrisy.

But please know that I'm not mocking thoughts or prayers. I've offered them myself to survivors and families after every national shooting tragedy. It's just that without action, thoughts and prayers are all but meaningless. Even the Bible, in James 2:17, says, "Faith by itself, if it is not accompanied by action, is dead." In this case, the action of lawmakers should be fighting for legislation that has been proved to reduce gun violence and save lives.

Managing the Downsides—Trolls, Flameouts, and Outright Mistakes

As useful as social media undoubtedly are, they also come with their own set of land mines, including a plethora of trolls. I've already talked about the hateful comments our Facebook posts have garnered; I was surprised to learn that civility is often even lower on Twitter. When I first started my personal Twitter handle, I was completely taken by surprise at just how mean-spirited many of the comments I got were. People would post images of my face Photoshopped onto a naked body. Or they would make up lies about my personal and professional lives, commenting on my marriage, my parenting, or my job history. In the beginning, I responded to these attacks, which is the worst thing you can do because it almost always turns into a pointless, never-ending online argument. Now I live by the mantra "Don't feed the trolls."

Most of the time, trolls don't care what you have to say; they live only to belittle you or to engage in a fight. But if you get trolled online and think the comments may have a thread of truth to them, do a little research. Weigh how many followers commenters have, check to see whether they're verified, and then read their profile bios to better understand what they care about. They're most likely gun extremists who will never agree with you, but if they volunteer in your organization or work on your same cause, or if they are influential or verified, you should take their comments seriously.

For the most part, the mean-spirited attacks on social media have become white noise to me; they hardly register anymore. As I mentioned, we have volunteers whose sole job is to block trolls; because your feed is the online equivalent of your house, you don't have to let hostile people in. Or as someone on Twitter once put it: this hellscape is optional. (There's also a "mute" function on Twitter, which, frankly, I don't find valuable; I just block and move on.)

You may get pushback for blocking trolls—I've blocked so many people that the hashtag #ImBlockedByShannonWatts once trended on Twitter. But if you know how to use them, social media are great at empowering you to take back the narrative. When that hashtag popped, I went and found some of the most offensive things people had written about me on Twitter and re-tweeted them using that same hashtag so my followers could see why I needed to use the block function so frequently. It helped me become part of the conversation instead of simply the butt of a joke, which was why the hashtag had been started in the first place—to troll me.

I also trended once after I tweeted, "You guys the NRA says I'm an unhinged wolf because I advocate for background checks on every gun sale." (Why do so many troll insults make no sense?) The NRA wrote back, "Looks like another

Shannon Watts tantrum is brewing" and included the hashtag #ShannonWattsTantrum. This was meant to be something its own supporters would run with. The only problem is, the NRA's base is made up of older men who don't know how to use social media very well. The hashtag took off, but only because my followers, allies, and Moms Demand Action volunteers ran with it. Emerge Colorado—a state chapter of the national organization whose board I am on and that offers training and support to women who want to run for elected office—tweeted, "If having a tantrum means you're a woman getting shit done, then bring it. #ShannonWattsTantrum." Social media also help you grow an online grassroots army that will not only support your cause, but also have your back.

Another way to fight back against trolls is to report them. As I was writing this chapter, Gavin McInnes, the founder of the far-right group the Proud Boys—the self-described "fraternal organization of Western chauvinists" that spews plenty of hate online—called me a "stupid bitch." I reported him to Twitter, and he was suspended. Social media are not completely lawless—each platform has guidelines on what is and isn't allowed as well as consequences for flouting those guidelines. By reporting offenders, you protect yourself and you make social media platforms safer spaces for everyone.

Of course, along the way on your social media journey, you'll make mistakes. You'll write something without nuance, share an article that turns out to be fake news, or slip up unintentionally. I did this myself after one of our annual conferences called Gun Sense University. I sent out a tweet paraphrasing a speech that Stacey Abrams, then–gubernatorial candidate for Georgia, had just given to our group. And I got it wrong. My tweet made it sound like she was anti-gun, which she's not—and neither are we. It may seem like a small distinction, but we go to great lengths to be clear that we are *not* anti-gun, we're anti–gun violence. We really are seeking common ground with gun-rights supporters, and my tweet had the potential to undo a lot of the mutual understanding that we've been able to create. I deleted the tweet as soon as I realized what I'd done, but it was too late—gun-rights extremists in Georgia had already taken screenshots of my tweet and were sharing them with their networks. Articles were written about it, and I was mortified.

When you make a misstep on social media, how you handle it dictates how much damage it can cause. If you apologize and explain yourself, you're much more likely to see it die down pretty quickly. But if you get defensive, blame someone else, or, worse, attack someone else, it's like pouring gasoline on a fire. It's worth it to do a little thinking ahead of how you'll handle it when something goes awry on social media, and if you're seeking to organize others in your efforts, to put codes of conduct in place as soon as possible if you don't have them already. Your opponents and trolls are just waiting for something to use against you, so take precautions to minimize the chances of giving that something to them.

Moms Demand Action created a formal social media policy for our volunteers to follow when posting to their state chapter Facebook pages. For example, we ask them to never engage in personal attacks and to stick to subjects that are related to gun violence. We have moderators for all of our Facebook pages—public and private—who help monitor conversations and address concerns or redirect negativity whenever possible.

On some of our Facebook pages where the conversation can be more sensitive, our moderators review individual comments before they are posted. Given that our subject area includes politics and violence, and that many of our audience members are survivors still dealing with trauma, our guidelines help ensure that we're providing a safe space for people to strategize about gun violence prevention.

It's not always easy to refrain from attacking someone, especially when dealing with online trolls. The conversations can inflame your emotions pretty quickly, and it's so tempting to fire back. But as Glennon Doyle, creator of the online community Momastery, says, "There are not two of you—Internet you and Real you. There is only one of you. If you aren't kind on the Internet, then you're not kind." That's just one more great reason to have an online policy about what you can and can't say as a representative of an organization, paid staff or volunteer. In the most heated moments, it helps to have a defined direction before you send off something you'll later (or, usually, pretty much immediately) regret.

Harness the Power of Social Media to Change Minds—and Lives

So much of the work that we accomplish via social media isn't the result of anything I've done—it's thanks to the hard work and big hearts of our volunteers. I want to let them share, in their own voices, some of the ways social media have helped them bolster our cause, create community action, and even enhance their own lives.

- **Find places to show up in real life.** Jane Hedeen, the leader of the Indiana Be SMART initiative—which educates parents about how to store guns safely and talk to their kids about gun use and suicide—uses social media to follow various public safety groups (such as police and fire departments) and community organizations (such as antiviolence groups and neighborhood groups) to suss out opportunities to showcase Be SMART. She says that social media have been "especially helpful in increasing the diversity of events we participate in and the diversity of communities we can partner with to spread the Be SMART message." Reba Holley, local group lead of the Mercer County, New Jersey, chapter, uses social media to find events where she can show up in her red Moms Demand Action T-shirt and spread the word about gun sense simply by being there. "I hand out a lot of business cards, and some people even join us," she says.
- **Break the ice.** I've heard from volunteers that interacting with others on social media makes it easier and less intimidating to attend their first in-person events. Jack Kohoutek, one of our many male volunteers, admits that "social media is where I do most of my relationship building; without it, I'd probably have had a few (at best) awkward conversations with Moms Demand Action volunteers in real life and maybe not come back." Kristi Cornett of our Tennessee chapters echoes that sentiment: "As someone who struggles with social anxiety, getting to know Moms Demand Action people online before I actually met them also helped me to keep showing up."
- **Feel less isolated.** Whatever cause you're passionate about, it may not seem like the people who are currently in your

life share your views, and that can feel awfully lonely. But there generally comes a point when it's more painful to not speak your truth—even if it means you alienate family and friends. That's what happened to Julia Chester of the Indiana chapter. When she started posting her views about gun safety, she lost touch with many family members and friends "because they just couldn't handle it." But what social media taketh away, they also giveth: "The friends I've gained have changed my life; they also gave me the emotional support that's required to stay in this movement." Jessica Pettigrew of the Colorado chapter has had a similar experience. She was nervous to re-post Moms Demand Action posts on her own page, but she says that more times than she can count, women (and men) whom she went to high school with have reached out privately to encourage her. "They say 'thank you,' or, 'Wow! You are saying what needs to be said.'" Social media make it easy to share your views and discover who else in your life belongs to your tribe.

- **Plant seeds in people's minds.** While social media move quickly, using them helps you play the long game—every post is a seed planted, and you never know when it will sprout. A member of our Arizona social media team, Kara Waite, reports that after the 2018 Parkland school shooting she received "heartfelt notes from friends who have seen me post time and again about gun violence and a) wanted to thank me and Moms Demand Action for our work, and b) wanted to be connected with a chapter." Kara fielded so many of those messages that she was able to match about a hundred people who reached out to her with a member of their state chapters of Moms Demand Action so they could hit the ground running.

- **Inspire others to do intimidating things.** Democracy may not be a spectator sport, but sometimes getting a picture of what it actually looks like makes it easier to dive in. Rachel Guglielmo of the Indiana chapter has found that "sharing pictures of ourselves having fun doing 'scary' things, like phone banking or canvassing, encourages people to try it out for themselves."
- **Be a lifeline to people in need.** Having a social media presence means that when people are upset and in need of support, they have a much better chance of finding solidarity with like-minded people. That's what happened for Pooja Mathur of our California chapter: "At my saddest moment, reading about Sandy Hook, as I cried in front of the computer, feeling helpless, Moms Demand Action appeared in my Facebook feed. I clicked 'Like.'" And now her local group regularly has one hundred people showing up for meetings—one hundred people who can understand that sadness when it crops up. Kristin Bell Gerke of our Florida chapter shares, "After every tragedy, I get messages from people who say, 'I know you're involved. What can I do? Where do I go from here?' People know where to turn when they feel powerless, and we can help them gain their power back." Our social media platforms have also helped many survivors find each other and realize they are not alone, as happened when Tony Cope, the social media lead of our North Carolina chapter, shared his thoughts on his own Facebook page about being a survivor of gun violence.

Much to his surprise, "I discovered three friends were also survivors. They said hearing my story helped them start telling their own stories."

- **Be findable by media.** When using social media like Twitter or Instagram, posting about the cause you care about along with a searchable, relevant hashtag will make it much easier for reporters to find you and interview you about your work. Becky Morgan of Missouri Moms Demand Action has had this experience, noting that "numerous reporters who probably wouldn't have found me have reached out to me for interviews via Twitter."
- **Rally the troops.** Social media platforms are also incredibly convenient for getting the word out to a large group of people whenever you need to summon a crowd. Kristi Cornett of the Tennessee chapter says this is how she's gotten others to show up at the statehouse to support some important gun-sense bills: "By being able to put out calls to action on social media, we packed the room with only a few hours' notice."
- **Keep an eye on your opposition.** Moms Demand Action's campaign to get Chipotle to change its open carry policies might never have happened if it hadn't been for Mindy David, a Texas chapter volunteer, who was browsing gun extremist social media pages when she found a photo of two armed men taking their guns into the restaurant. She shared the photo in a private group of her local Moms Demand Action chapter—it was then shared with our social media team, who posted it along with our #BurritosNotBullets hashtag. It went viral. The image of the two men became ubiquitous on social media and was highlighted on Stephen Colbert's

The Colbert Report and on Bill Maher's HBO show *Real Time with Bill Maher.* Within just a few days, Chipotle changed its gun policy to disallow open carry inside its stores. That one photo created such a big ripple, that we realized we had a strategy on our hands. A team of Moms Demand Action volunteers from Nebraska, New York, Virginia, and more went looking for others, and their research played an important role in our campaigns to influence Sonic, Target, Chili's, and other businesses.
- **Support one another in times of need.** The sharing we do on social media isn't just about our work; we also talk frankly about when we're feeling down, which then helps us get the support we need just when we need it. As an example, Jennifer Rosen Heinz of the Wisconsin chapter was feeling super burned out after Parkland and posted something about it on Facebook. "An hour and a half later," she says, "Mindy Rice, a fellow volunteer, was on my doorstep with a home-baked chocolate cake." Jennifer emphasizes that social media help us display our commitment to one another and show that "Moms are ride or die."

As I've said, I know social media can be a time suck, and it may be a little frightening to put your views out there in such a public way. But ultimately, social media platforms provide a powerful bullhorn. I hope you'll be inspired to pick up one, or two, or more of them and start using them to find your people and make the changes you want to see in this world. Moms Demand Action is living proof that your people and your power are only a click away.

6

Tap into the Priceless Power of Volunteers

The word *volunteer* sometimes gets a bad rap, conjuring up images of bumbling, disorganized, and well-meaning but naive people trying, and failing, to get things done—like the characters in a Christopher Guest mockumentary. And in our capitalist society, Money = Value, and unpaid work is often not considered "real" work, whether that's volunteering or taking care of kids.

Both volunteers and moms—and perhaps to a greater extent, moms who volunteer—often don't get enough respect for the work they do. But being a volunteer in fact gives us enormous credibility *because* we don't get paid. We have no financial skin in the game. Contrast that with NRA leaders and lobbyists, who are driven by money. Their spokespeople and lobbyists are pulling down fat salaries to help gun manufacturers sell more guns. In 2015 executive vice president of the NRA Wayne LaPierre took home $5.1 million; I took home $0. Whose motivations do you trust?

From the very beginning, I vowed not to take any compensation for my work with Moms Demand Action. That's because I want to do this work as an advocate who is driven by passion and commitment, not money—and because I want our volunteers to know that I am just another member of their army, even if I often serve as the tip of the spear. Of course, I'm lucky and privileged that my husband is able to support our family without my needing to earn an income, and I'm honored to be able to put that good fortune to use doing such important work. This makes it incredibly ironic when the NRA accuses Moms Demand Action volunteers of being paid to show up and protest. But really, we shouldn't be surprised—it's a classic hallmark of manipulation and narcissism to accuse your opponent of doing the exact thing that you yourself are doing.

Moms Demand Action volunteers are motivated by something much more powerful than just fear: love. Love drives us to protect kids, women, people of color, trans and gay people—everyone who is vulnerable to gun violence, which is everyone. Protecting our families and our communities is something we can't *not* do. It's absolutely authentic to who we are at our core, and it puts us in stark contrast to the cold calculations and manipulations of the gun-manufacturing industry.

As volunteers—and as moms—we provide a counterbalance to the NRA in so many ways beyond the most obvious (we're primarily women; they're primarily men). These differences are what give us our unique power:

- We care about protecting children; the NRA cares about protecting profits.
- We believe in dispelling myths and making decisions based

on data that illuminate the most powerful ways to save lives; the NRA traffics in misinformation and fear-mongering anecdotes.

- We stand for collaboration and partnership—true hallmarks of the feminine archetype; the NRA is about bullying, posturing, and attacking—the classic components of toxic masculinity.
- We have a growing base of engaged volunteers who show up and aren't afraid of heavy lifting and who have the know-how to leverage modern tools such as social media; the NRA's base is primarily older white men who generally *don't* show up (we continually outnumber NRA members in hearing rooms) and are either apathetic about or flummoxed by social media.
- Our biggest weapon is our grassroots army of volunteers who show up when they're needed, for however long they're needed; the NRA's is its annual budget of more than $300 million.
- We're on the right side of history; the NRA is clinging to a past that no longer exists.
- We seek to bring people together, from red states and blue states, within families, and among different populations; the NRA seeks to divide by standing down in the face of propaganda that promotes violence against "others."

Often, we're also a visual counterbalance to the NRA in the halls and hearing rooms of statehouses across the country. Gun-sense candidate Jennifer Wexton, who is now the new congresswoman for the tenth district of Virginia (the literal home of the NRA), served in the Virginia State Senate from 2014 to 2018. "When I first got to the state senate," she says, "every Martin Luther King Jr. Day was gun lobby day, when the Republicans would bring all the gun bills before the Courts of Justice Committee—and they'd kill all the gun safety bills and move along all the bills that put more guns into the hands of more people in more places."

Aside from the terrible irony of using as an occasion to loosen gun laws a national holiday dedicated to a civil rights leader who committed his life to nonviolence and was assassinated by a gunman, Wexton says these lobby days were made all the more upsetting by the presence of numerous gun advocates "open carrying with great displays of armed force." And to add insult to injury, no one was showing up in any organized fashion to advocate for strengthening gun laws. At least not until 2015, when the Virginia Moms Demand Action chapter was organized enough to send volunteers to the statehouse to make a strong showing—and our numbers have only increased every year.

Our very presence, wearing our un-ignorable red T-shirts, provides a hugely important display of a differing point of view; we're a visual reminder of all the Americans—more than 90 percent—who support keeping guns away from dangerous people.[1] As Wexton says: "Having Moms Demand Action volunteers start to show up made a huge difference—it raised public awareness and it boosted the morale of those of us who had been fighting this fight in the past. The support of Moms Demand Action is not just a shield—it's a sword."

Before our volunteer army existed, entirely too many bad

gun bills moved through statehouses like hot knives through butter. Now Moms Demand Action volunteers regularly outnumber gun-rights advocates by an order of magnitude, and our presence plays a huge role in 90 percent of bad gun bills being defeated in statehouses across the country each year.

When we volunteers show up in large numbers, and when we start talking about the issues we care so much about that frankly we can't *not* talk about them, we do something that no politician can do on his or her own: we sway public opinion. This is a big deal. As Abraham Lincoln said, "public sentiment is everything. With public sentiment, nothing can fail; without it nothing can succeed." And this influence helps our lawmakers pass the laws that will keep our kids and families safe. Speaker of the House Nancy Pelosi puts it this way: "Knowing that moms are out there relentlessly shaping public sentiment toward commonsense efforts to prevent gun violence gives me the power to negotiate." Lawmakers can deliver the laws if we deliver the voters.

Overcome Doubts

One of the NRA's biggest criticisms of me—aside from the fact that I'm a woman—is that I'm "Astroturf," which is to say, "fake." Because I graduated from college and had a robust career before staying home with my kids, and because I occasionally took on some consulting gigs when I wasn't employed full time, I'm accused of not being a "real" stay-at-home mom—which many NRA members perceive as being someone whose biggest achievements are cooking and cleaning. A "real" stay-at-home mom—and probably any woman—couldn't possibly create an organization as successful as Moms Demand Action. That's why the NRA refers to me as a "Bloomberg whore" or "puppet"—in part because Moms Demand Action partnered with Mayors Against Illegal Guns (which was started by Michael Bloomberg) to form Everytown for Gun Safety, but also because there just has to be a man behind anything so effective.

It's a sad but true reality that you can almost always expect a group led by older white men to be skeptical of women's ability to do great things. Yet there's a way we spin a version of this narrative in our own heads when we think about getting involved in the effort to effect change. It shows up in thoughts like "Who am I to get out there and think I can do something to solve this?"

A crucial component of being an effective volunteer is not talking yourself out of the impulse to do something. Because that's what we so often do—we see a Facebook post about an issue we care about and we think, "I have to do something about this." But then that little voice of doubt chimes in and says: "I don't have the bandwidth to get involved. What can one person really do anyway? Someone else can take care of this better than I can."

I can't tell you anything that will 100 percent eradicate that voice of doubt—it's probably ingrained in all of us. So we may not be able to make that self-doubt, that is, our ego, go entirely away, but we *can* learn how to pay less attention to it. I think of dealing with the ego as being similar to the way you deal with that well-meaning relative who always makes subtle (or not so subtle) judgey comments at the Thanksgiving table: you view the comment in your mind as a cloud and watch it float away. Whatever tactic works for you to turn down the volume on that voice, use it. Because truly, that voice is the only thing standing between your desire to make change and your ability to do so. And the world needs you to stop giving that voice so much credence.

As a woman, and as a mom, you are as powerful and as awe-inspiring as lightning; the question, then, is how to contain and then direct that power so it doesn't disappear in a flash.

How We Bottled Lightning

I frequently describe the beginnings of Moms Demand Action as catching lightning in a bottle—that's the only accurate way to describe going from seventy-five personal Facebook friends to six million supporters and hundreds of thousands of volunteers in just five years. Many organizations would give anything for that kind of grassroots power, and I'm incredibly grateful we tapped into it and harnessed it.

As I've said, it all started with a post to my personal Facebook page that said I wanted to start a conversation with other moms about gun violence. One of my seventy-five friends saw it—ironically, he was a man and a gun owner I barely knew whom I'd met through another friend—and connected me to a woman in Brooklyn. This woman was all over our mission and went on to gather a big group of friends who also lived in New York City to organize a march across the Brooklyn Bridge on Martin Luther King Jr. Day, which then drew in to our cause a lot of other people. Simultaneously, some women from the San Francisco Bay Area reached out to me and offered their help—and became instrumental in helping us organize volunteers into chapters.

The strangers I met through this one Facebook post ended up doing a lot of the behind-the-scenes, time-consuming work of birthing an organization—navigating the process to become a nonprofit organization, creating basic materials and tools to bring new volunteers on board, and creating individual Facebook pages for each state chapter. I didn't realize it at the time, but I'd stumbled on a formula for going from outraged to engaged: I got connected to others who wanted to help, we divvied up the duties, and then we started taking one action after the other. We stayed in contact with each other and collaborated, but we also each ran with whatever project we claimed. There was no micromanaging. Everything else snowballed from there.

This is the exact same formula that we still use to this day to continue growing our volunteer base by reducing the possibility that doubt will derail someone's desire to make a difference: we quickly connect people to other people in their community, we encourage them to take action, and we empower them to take ownership.

Welcoming prospective volunteers into an organization and giving them something to do or a place to go is essential when cultivating your army. In fact, many people have told me that the reason they joined Moms Demand Action was because we were the only organization to return their call. When someone has that moment of knowing that they want to get involved to right a wrong, we want to capture that momentum before the voice of doubt kicks in, or the news cycle moves on, or daily realities cause that person to lose interest. That's why we make it a point to engage anyone who expresses an interest in volunteering as quickly as possible in four ways:

1. **Personal welcome.** Someone from our robust welcome team personally calls each person who signs up to volunteer and welcomes them into the organization. It's such a simple thing, but it makes a huge difference in creating engaged volunteers who feel valued. To give you an idea of just how much energy we devote to this effort, in the first half of 2018 alone, our welcome team made more than eight thousand of these calls.
2. **Invitation to an event.** After that, a state membership lead and her team take over and make it a point to invite new volunteers to show up at an actual event as soon as possible. We want to get our eyes on you so that we can start to get to know you and you can start to get to know us and feel like you've found a welcoming home. That event may be a monthly chapter meeting, or, if your state legislature is in session, we may send you right into your statehouse for a gun bill hearing.

   I've heard from so many people that attending their first meeting really cemented their dedication to becoming a Moms Demand Action volunteer. A typical tweet from first-timers sounds something like this: "Just went to my first Moms Demand Action meeting and no one gets shit done like a bunch of focused, angry moms!" I recently heard from a man who had worked in our New York City Everytown office for years. After moving on to a job in California, he was still devoted to the cause. He said, "I'm embarrassed to say I just went to my first Moms Demand Action meeting last night and it was amazing. There was laughter, there were tears, there was cake."

Once you've attended a Moms Demand Action meeting, you feel like you've found your people, especially if you live somewhere where you feel isolated because you're outnumbered by people who don't share your views. And it's not by accident—we make our meetings inviting and warm, as well as incredibly productive and efficient. I can't take credit for that, though—that's due to the time and talents of our chapter leaders, who have really taken these meetings and run with them. This is another key piece of our "keep volunteers engaged and fulfilled" strategy: let them lead, and trust they'll hit it out of the park.

3. **Take on a task.** As part of our strategy of empowering volunteers to lead, after you attend your first event we encourage you to take on a task that is dependent on what you want to do—you could work with the legislative lead to help support a good gun bill or block a bad gun bill. Or you could work on our Be SMART initiative to prevent unintentional deaths from kids getting ahold of unsecured firearms. Or you could help plan an event or become part of the welcome team.
4. **Become a leader.** Eventually, we'll ask you to take on a leadership position for your local group or state chapter. And not just because we want to share the workload, although, of course, we do. It's because we know that volunteers are more likely to stay in the fight if they feel personally invested in the cause and rewarded for that investment. It's important for volunteers to feel they have skin in the game,

and Moms Demand Action benefits from the unique talents each of our volunteers brings to the table.

Because I founded Moms Demand Action and act as a spokesperson, people often thank me for my work. But it's not *my* work that's made our organization successful—it's all of *our* work collectively across the country that has helped us grow and win. After all, there is no *I* in *mom*. Everything we do, we do collectively.

A principle that inspires me greatly and informs our culture of collectivism is Shine Theory—a term that was coined by Ann Friedman and Aminatou Sow, co-hosts of the podcast *Call Your Girlfriend*, that describes "a commitment to collaborating with rather than competing against other people—especially women." In other words, when we let women shine, we all benefit from the glow.

It's because of Shine Theory that Moms Demand Action won't stand for women doing only the behind-the-scenes work, which so often happens in other grassroots organizations. Our volunteers get to do everything, and that includes setting the priorities and getting the glory. Adhering to Shine Theory not only makes the work more fun and fulfilling, it also encourages greater performance. Our culture of collaboration is what helps us get so many wins. And winning is very motivating!

Of course, it isn't only wins that we celebrate, because, as I've said, we very much believe in the value of losing forward. So we celebrate our losses too—especially when those losses hurt. Nothing helps you move through disappointment better than having an army of women pick you up, dust you off, recognize your hard work, and remind you just how many soul sisters you have supporting you. As they say, Don't let the bastards grind you down—and having the support of other women fighting by your side will help keep your spirits and your strength up.

What to Do When Volunteers Leave

Although we help boost each other up when we get knocked down by external forces, I have to acknowledge the ups and downs that can arise internally, too. Inevitably, we periodically do have unhappy volunteers—or groups of volunteers. And sometimes that unhappiness comes out very publicly in a big online explosion.

As helpful as social media can be for a grassroots organization, they also can have a downside. Primarily, social media make it easy—sometimes too easy—to share with the world how you feel about a personal issue instead of sharing it with the person you have the issue with. Sometimes volunteers will leave the organization. And chances are that their departure will get shared on social media. This is a new nuance to organizing online that there isn't much expertise on how to manage.

Moms Demand Action has a policy that urges volunteers to communicate any complaints or concerns privately before sharing them publicly. We always aim to remember that we're a family, and we try to handle internal disagreements the same way a family would—in other words, we don't want to air our dirty laundry in public. It's also a priority for us to treat one another with kindness. As most of us have experienced in this internet age, it's all too easy to be unkind or critical online when you're not looking someone in the eye, or even talking to them on the phone.

One way to ward off the worst of the Facebook flameouts during times of controversy is to hold private forums where people can express their concerns and, in turn, you can update them on your thought process. We've held many conference calls to do just that. Whatever format you choose, it's better to have them earlier rather than later, even if you haven't fully figured out an official stance yet. You can share your current thinking and ask for feedback. Ignoring a problem or putting it on the back burner is a surefire way to turn an issue into a crisis.

All that being said, it's impossible—and wrong—to try to shut down dissent. Every organization must allow debate; but it's important to set up protocols that help keep the debate from veering into personal or unfounded attacks.

Inevitably, there will be instances when our volunteers will disagree with where Moms Demand Action is headed as an organization—to the point where they may leave in a minor, or major, revolt. We've experienced this a few times, but the first and the biggest time was after we decided to become a part of Everytown for Gun Safety.

To set the stage on how that development came about, it's important to know that during our first year of existence, we raised about $200,000. That seems like a lot, and for a brand-new organization it was, but it wasn't nearly enough to grow the army we knew we needed to do the work we wanted to do.

We used that money to fund a few organizing manager positions, to pay a monthly retainer to a PR firm, to buy swag, and to hire a fundraiser. The more people our fundraiser met with, the more it became clear that people don't give money to organizations run by women they've never heard of. We got the message again and again that we needed an influential partner with name recognition, and I started to see that this feedback was accurate.

I felt very strongly that we would not survive without more resources. Although our fundraising efforts were bringing in some money, it wasn't enough. (It's really unfortunate that it is so difficult for grassroots movements to raise funds.) We were living month to month, and we often didn't know whether we'd have enough money to pay people. Still, we knew what we'd created was valuable; I certainly didn't want to just give it away. I wanted to enter into a marriage that would provide benefits to both parties.

So I started reaching out to all kinds of advocacy organizations—many of them related to gun safety, but some not—to explore a potential partnership. I said to them, "We have this grassroots army to give you, what can you give us in exchange?" Ultimately, we had the most in common with Mayors Against Illegal Guns (which is now part of Everytown); they were also pivoting to states and corporations, and they had a lot of human and financial resources.

During the summer of 2013, I met with Mark Glaze, the executive director of Mayors Against Illegal Guns at the time. We met in Montana and rode a ski gondola up and down the mountain a million times talking over how it could all work. We finalized the deal internally at the end of 2013, and Michael Bloomberg and I announced our partnership formally on *The Today Show* in April 2014 (where I also got to meet Cyndi

Lauper in the greenroom; oddly, she thought I was Mike's bodyguard—I guess she could see my inner badass shining through!).

Even though partnering with Everytown relieved Moms Demand Action from living month to month, we still encourage all volunteers to fundraise for us and we train them on how to do so. A tactic that's been very effective—a birthday fundraising ask on Facebook—combines the power of volunteers with the power of social media.

It was Debbie Weir, the former CEO of MADD who became our managing director in 2017, who told us we had to teach our members how to raise their own money in addition to the budget they receive from Everytown. This is because there will always be things to spend money on that aren't in the budget; and if you don't know how to raise your own money from the very beginning, there won't be time for you to weather the learning curve when something urgent comes up that requires extra funds. As powerful as having a large volunteer base is, you still need money to help you have more reach and more impact, so don't fall prey to the desire for someone else to take care of it for you.

Recovering from Growing Pains

Our partnership with Everytown turbocharged our effectiveness and growth, but it also ushered in a wave of many of our original volunteers leaving. Granted, most of our volunteers were thrilled because they knew we now had access to important resources—mainly money, research, and clout—that would help us win more and last longer. But some were put off by what they felt was a loss of freedom. We were implementing a new strategy and structure, and that change was too much for some volunteers who were used to having more leeway. And other volunteers were just plain ol' tired after working hard for more than a year, and they felt like the organization was in a good place and so they stepped away.

While you're never going to get 100 percent agreement in any organization, particularly one that focuses on life or death issues, this was a painful experience. I had disappointed people who had given their time and talents to our organization, and I lost friends I'd been in the trenches with since the inception of Moms Demand Action. But in retrospect, the change was necessary to facilitate new growth, even though it was not fun to live through.

To regain our momentum, we turned to the list of people who had contacted us after Sandy Hook and the Manchin-Toomey loss and wanted to get involved; now that we had the volunteers and staff, we could effectively welcome and enroll them. We also developed a strategic outreach plan of hosting talks across the country to spread our message and invite new folks to join us. It took about six months to really hit our stride again, as these efforts took time, but it was a fundamental part of helping us grow into the force we are today.

Welcoming an Influx of New Volunteers

We have the bittersweet advantage of working on a cause that people are forced to reckon with repeatedly—we see a bump in volunteers at the start of each school year, when kids as young as preschoolers start having to do lockdown drills and their parents begin to worry that the next school shooting will be in their town. People also come to us when there's a shooting in their family, in their community, or after a national shooting tragedy—we took in hundreds of thousands of new volunteers after the shooting in Parkland, Florida, alone.

Working on an issue that rears its head so frequently means that we always have a steady stream of new people coming onboard. In a macabre way, this is good, because having a consistent flow of new people joining is a sign of health: it means your organization's message is still compelling, and it brings in new energy and new ideas. Yet it's also challenging because those new ideas may require you to change and evolve—two things that humans instinctively resist.

The way we've organized enables us to bring volunteers onboard quickly and effectively. Our organizational model is based on the "snowflake" model, developed by the political organizer Marshall Ganz and used by, among others, the Sierra Club (it's a pretty ironic name given other contemporary uses of the term "snowflake"). In the snowflake model, each state has a volunteer chapter leader in the center with a circle of volunteer positions supporting her. The volunteers are responsible for data entry, event planning, and everything in between. Yet, because each snowflake is different and so is each state, the particulars of those volunteers around the center can vary. Ideally, each chapter leader holds her position for two years while cultivating a new leader from her snowflake to whom she can pass on her responsibilities when she is ready to move on. We also created regional manager roles, which are paid positions, to help manage the state chapters.

This organizing model is built to grow and change along with your organization. As proof, we have changed and adapted our leadership structure multiple times since we started. As I mentioned above, after the shooting tragedy in Parkland, Florida, Moms Demand Action chapters across the country absorbed nearly two hundred thousand volunteers in just a few weeks. Meetings that typically included two dozen people suddenly had to accommodate hundreds and even thousands of new volunteers. It was still a ton of work to reach out to everyone and to adapt to our new reality, but because we had the structure in place, we had places to put people. And it worked—we retained a huge percentage of the volunteers who sought us out during that time.

While volunteering is crucial for organizational growth, it's also a powerful mechanism for personal growth. Volunteering gives you opportunities to unleash and channel your inner badass, as I covered in Chapter 3. That's a positive development, but it can also send shockwaves into other parts of your life—such as your marriage and your parenting—that might at first feel destructive, especially if you aren't on the lookout for the positive changes that they also leave in their wake.

Navigating the Changes Volunteering Can Make to Your Marriage

It is absolutely true that volunteering can help you recognize your own power. But it is just as true that as you start to own that power, ripples will move out into your life that can affect all your relationships, including your marriage. It certainly changed mine.

Just to give you the backstory: I got married to my first husband right out of college when I was twenty-three. By the time I was twenty-nine, I had three kids and was the primary breadwinner for our family. By the time I reached my mid-thirties, my then-husband and I realized that we'd grown up and become two completely different people. We divorced soon after, although we remain committed co-parents—he lives near me and our kids, and we share custody.

When I started Moms Demand Action, I had been married to my husband John for only five years. We were still in what we called "the merge": head over heels in love with all of the accompanying hormones and neurochemicals that keep you glued to one another's hips. We were just coming out of that phase when Sandy Hook happened. We went from the honeymoon period straight into a tornado—one that we hadn't had a chance to plan for.

John was supportive and told me that night I created the first iteration of Moms Demand Action that he thought my Facebook group would be "really big." He gamely stepped up and started doing even more of the cooking than he was already doing, as well as the shopping, the childcare, and the chauffeuring for our now-five kids (he has two daughters of his own). But the fact that I transformed basically overnight from stay-at-home mom into more-than-full-time volunteer threw us into turmoil.

John is twelve years older than I am, so sometimes we've had to marry our two different ideas of what a mom is and what a marriage is. Luckily, because we'd both been married before, we'd made a commitment when we got together to go to marriage counseling every two weeks, no matter what. Through that, we explored our marriage as a container—one that has to be able to hold each individual as well as two individuals as a couple. You have to be able to have your own individual lives and to exist together, and the happier you are as an individual, the happier you'll be as a couple. When you have that container, nearly anything can happen, and your bond will survive and thrive.

We've spent a lot of energy trying to figure out how to stay connected at the same time that I'm doing this work that I find so fulfilling—and that frequently requires me to travel. We had to figure out how to avoid his feeling lonely and how to avoid my feeling constricted—because neither feeling is tolerable. We've worked hard to construct a container that allows us to be connected and able to pursue our respective passions. We went through a fair amount of growing pains, but they were worth it.

The other important piece counseling has helped us put in perspective is that marriage is a long game. Now, six years into Moms Demand Action and eleven years into our marriage, I'm not so in-the-trenches as a volunteer (more evidence that sharing the workload is key for longevity and happiness). I've had time to go to graduate school and write this book. And now he's taken a job that requires him to commute from our home in Boulder, Colorado, to San Francisco Monday through Thursday, so I'm back in the role of primary parent. (The kids and I eat out when he's gone—the folks at Chipotle know us by name.) It's his turn to pursue something wholeheartedly that he finds fulfilling, and it's my turn to be happy and excited for him as he does a job that he loves. Had Moms Demand Action not happened, I probably wouldn't have known how to support him in that way yet. He taught me.

I'm incredibly proud of the fact that John and I have been able to strengthen our marriage even amid the growth of Moms Demand Action. I also have to admit that I love seeing him be proud of me—even when my work can sometimes encroach on our private life. For example, we recently went to Maui for vacation, and someone from the local chapter found out we'd be there and asked us to come to an event. It wasn't necessarily something John wanted to do, especially because there was a lot of traffic and it took us an hour to get there. But once we were there and I gave a speech, I could see him beaming with pride the whole time.

The most important thing to keep in mind is that you want to always communicate with your partner not only about the new logistics your volunteerism will create in your relationship (you may not be on hand to cook the dinner or watch the kids as much as you were before), but also about how much it matters to you. Ultimately, your partner wants you to be happy, and if volunteering contributes to that, you'll find a way to make it work. This may happen through trial and error, but I believe that there is always a way for both partners to get their primary needs met.

Sharing Your Kids and Your Cause

As I mentioned, before Moms Demand Action, I was a classic helicopter parent; I was super involved in all the kids' school and sporting events, the homework, the friend drama . . . I was in it to win it. Practically overnight, my helicopter that had been hovering over my children started flying all over the country.

The kids told me a few times along the way that they were sad I was gone so much. It was certainly a switch for them. I'd been at home with them for the previous five years, and they'd gotten accustomed to my being there when they needed anything. But it wasn't as big a shock as it could have been, because I had worked their whole lives as a corporate communications executive up until John and I got married.

My work with Moms Demand Action was a blessing in disguise because it forced me to not be so tightly involved with them—it was better in the long run for my kids and for me, as it gave us all some independence. My oldest daughter, Abby, in particular felt a palpable sense of relief that I was no longer always on her about homework and grades. And the kids still had two other parents (John and my ex-husband)—a situation that has actually been quite beneficial, because having an extra parent means that a lot less falls through the cracks. Thanks in large part to this setup, they were always well tended to, even when I was on the road. And really, it's not a bad thing for men to have to step up in the parenting arena. When Emma got an award for being on the soccer team her senior year and I was away at a Moms Demand Action event, her dad and stepdad walked her across the field together—that's a special experience none of them would have had if I'd been around.

Even my youngest, Sam, whose nonresponse to the Sandy Hook shooting played such a big role in inspiring me to create Moms Demand Action, has gotten on board in a big way.

He's the same age as the Parkland kids and voted for the first time in the 2018 midterm elections. He proudly took part in the National School Walkout for gun safety in April 2018 and supports my work with Moms Demand Action. There have been periodic pangs of unhappiness, but it's important to me as a parent to show my kids that women can do whatever they want in life. The trick is to balance that with tangible reminders that they're always my priority.

That's not always easy; I've certainly had major parenting challenges in the midst of running Moms Demand Action, and at times I have felt the conflicting pull of having both a family and an outside passion. One of the hardest challenges was Emma's eating disorder. Devastatingly, she was the victim of acquaintance rape in the summer before she left for college, although she didn't tell us until she came home for Christmas break. By that time, she had begun to develop anorexia, but we didn't recognize the signs. We got her treatment for the trauma, but not for the eating disorder. After she returned to school, it got worse.

The following Christmas break, we realized Emma had a serious problem. She ended up dropping out of school, being hospitalized, and then going into a long-term treatment center in Denver. After she finished treatment, she transferred to a new college, but she relapsed before the quarter was out.

Several times during those harrowing months, I'd be on the way to the airport to attend a Moms Demand Action event, when I'd get a call from Emma and realize the place I needed to be was with her. It was hard to let people down, but it was non-negotiable: I had to be there for my daughter when she needed my support the most. It was also a gift for me to realize that the world wouldn't end if I wasn't available to my colleagues every moment of every day—they've got my back, and our movement is in great hands when I'm not there.

On the other hand, I might have been overwhelmed by Emma's eating disorder if I hadn't had Moms Demand Action. I'm sure most moms can relate—it's our instinct to blame ourselves for our children's misfortunes, and to allow ourselves to be only as happy as our unhappiest child.

I've tweeted about Emma's eating disorder, and some people have told me that it's too private a topic to discuss so publicly. I understand why they might feel that way—there's a lot of shame around eating disorders and mental illness. But Emma herself talks about it openly. Eating disorders are a disease, and talking about her anorexia is no different from talking about any other illness. If anything, shame is what leads to an eating disorder, and being able to talk about it is a big part of Emma's recovery.

She also had plenty of extra moms—or I should say, Moms—pulling for her and supporting her during her recovery. Emma, who turned sixteen just days after I founded Moms Demand Action, has been an involved member each step of the way. I've taken her and all of our kids to Moms Demand Action events because it's important for them to see what I'm working on all day, every day, and for them to experience the empowerment that comes from banding together with others to fight for something you value. Emma has also gotten involved in the pride movement—it's been incredibly poignant and moving to see such a visible reminder of why I do this work, because I want everyone's children to be able to do the things they care about without being vulnerable to gun violence.

All that being said, I get that because my kids were older, they didn't need me as much as they would have if they had been toddlers, preschoolers, or even in elementary school. You may be at a point of your parenting journey where you can't throw yourself wholeheartedly into a volunteer position. Maybe you don't even have kids but you have a consuming job and a full life and don't have a lot of time to offer. Just remember that you don't have to get nearly as involved as I did!

You can do plenty of things in little slivers of time, like calling lawmakers while the kids nap or sharing messages about the causes you care about on social media during your commute. They all add up, and they all matter. Those small habits of activism that you build now will make it that much easier for you to participate in a bigger way when your life changes—and it will change, eventually. It always does.

Why You Need to Say "Yes" to That Impulse to Volunteer

In this chapter I've talked about how doubt can keep you from saying "yes" to getting involved in a campaign for change, but there's one aspect of that doubt that I haven't talked about yet that is vital to address. And that's the feeling that it doesn't matter what anybody does; nothing will ever change. This is a sentiment captured in a 2015 tweet that's often re-tweeted around the December remembrances of the Sandy Hook school shooting that the British journalist Dan Hodges wrote: "In retrospect Sandy Hook marked the end of the US gun control debate. Once America decided killing children was bearable, it was over."

I would like to pinch the fat part of Dan Hodges's arm. His comment is so cynical and *so untrue.* This is America. And that means this is a democracy—government by the people and for the people. Without your direct involvement, democracy does not work. You can't just sit on the sidelines and say, "Wow, this is horrible." You have to get involved.

Every day I witness more and more proof that when you get involved, things *do* change. Best of all, our country gives us a framework for getting involved, by voting, by communicating with lawmakers and corporations, and even by running for office. The tools are there; it's up to each of us to use them—to use our voices and our votes. That means it's up to you.

You don't have to be rich, or powerful, or backed by a rich and powerful man to be effective. I wasn't anybody special when I decided that the current state of affairs was wrong and that I had to do something. And all the women who stepped up to help me weren't special either.

You are the perfect person to follow that inspiration you're feeling to volunteer for something you care about. We, your soul sisters, will be right by your side, helping you shine when you do.

7

Be Seen

On February 1, 2018, state representative Robert Byrd from the very red state of Tennessee introduced a bill that would allow teachers and school staff to carry a concealed weapon on school premises.

Nashville Moms Demand Action volunteers were having none of it.

When they found out that the bill would be heard by the House Civil Justice Committee on February 14, they made a plan to show up at the statehouse wearing their red T-shirts to testify against the bill. About ninety volunteers attended. While the committee was in session, the news broke that a school shooting was simultaneously happening in Parkland, Florida. In the wake of the horrendous news, the committee members decided to wait to vote on the bill until the next committee meeting; Moms Demand Action volunteers channeled their grief and anger over the shooting into fuel for the next round.

The Tennessee chapter leads then planned a big advocacy day for March 7, where six hundred Moms Demand Action volunteers showed up at the statehouse to meet with legislators—although the Civil Justice Committee opted not to convene that day. It ended up meeting on March 20, instead, a day for which we had not coordinated a large contingent to be on hand. At that meeting, members voted to pass the bill along to the next committee that needed to review it—the Education Administration and Planning Committee. "We were really nervous when we saw how easily the bill made it out of that [first] committee," recalls Kristi Cornett, one of the chapter leads. The Tennessee team knew they would have to up their game.

As soon as they found out the date of the next committee meeting, "We put out a call to volunteers from all over the state and basically begged them to attend—we had people who left their houses at 4 a.m. to make it to the statehouse on time," Kristi says. On that April day, forty Moms Demand Action volunteers packed the hearing room. The whole left side of the room was a sea of red shirts—some supporters even had to wait outside in the hall because of fire code restrictions.

The mostly Republican (male) lawmakers who sat on that committee were clearly taken aback; as they looked out at the crowd with wide eyes, Kristi says you could almost hear their hard swallows of surprise. Even though the numbers were clearly on the side of Moms Demand Action—there were only two NRA representatives in attendance—the volunteers were still nervous.

"Heading in, we really didn't think we had the votes we needed," Kristi said. Yet they did, when three Republican lawmakers changed their minds and voted to kill the bill. The final vote was seven to five.

To be clear, Tennessee volunteers had made hundreds of phone calls to those legislators too, and they had already shown up en masse twice before—once for the Civil Justice Committee meeting and another time for the advocacy day. But something about having all those red shirts in the room on the day of the vote changed the outcome.

A lot of ingredients go into successful change-making, but one you can't discount is branding. Having an instantly identifiable presence helps you build that key component that ultimately effects change: influence.

What the Heck Is a Brand, and Why Should You Care?

A brand is fancy marketing lingo for how you (because yes, an individual absolutely is a brand too), your product, company, or organization is perceived by others. It is composed of things that are tangible—such as your logo, your colors, your font choices, and your messaging—as well as things that aren't—your voice and your values. The sum total of these elements can, when done thoughtfully, add up to a brand that draws people in and engenders a sense of loyalty and respect—even before someone has directly interacted with you.

I learned the importance of cultivating a brand in the corporate world when I worked in marketing and communications for Fortune 500 companies like General Electric and Anthem. I knew from those experiences that good branding increases the value of a company, provides employees with direction and motivation, and helps generate new customers. The same is true in the nonprofit world: a strong brand helps enhance the credibility of an organization, gives its volunteers direction and motivation, and helps attract new volunteers. That's why Moms Demand Action has worked so hard from the very beginning to create a brand that makes women feel empowered as they go toe-to-toe with one of the most powerful lobbies in the nation, and conveys to everyone who hears our name that we mean business.

As the Tennessee story so aptly captures, our brand—with our instantly recognizable red T-shirts, tell-it-like-it-is name, and clear messages—helps us effect change.

How did we cultivate such an influential brand? We followed a few key principles.

Branding Principle No. 1: Know Your Audience

When building a brand, it's important to understand who your target audience is—in other words, who will buy what you're selling, whether what you're selling is something intangible, like culture change, or something you can hold in your hand, like a bar of soap. When I started Moms Demand Action, it was because an online search of gun violence prevention organizations didn't yield any that spoke directly to me as an American mom. I wanted to be a part of an army of mothers across the country who would fight for the safety of their families and communities. And, based on the response to my Facebook page, I knew I was onto something—many of America's eighty million moms felt the same way.

One California volunteer said that after the May 2014 Isla Vista shooting near the University of California–Santa Barbara, "I was so distraught. I had two sons in college at the time. I Googled 'moms against gun violence' to see what would come up. Moms Demand Action, which had been founded two years prior, was the first website that popped up in my random search. I joined immediately and have been devoted to this amazing movement ever since. It was the word 'moms' that was so important to me."

That's why everything we did in the early days of Moms Demand Action was aimed directly at American moms, from our organization's name to the stories we told on our social media platforms to the look and feel of our graphics.

And although I wanted to appeal to moms, I didn't want to create a brand that represented the stereotype of moms that presents them as soft and doting, or frazzled and exhausted, or constantly cleaning and cooking (as many of the commercials geared toward mothers still tend to show moms). "Mommies Demand Action" simply wouldn't have cut it.

I wanted to create a brand that represented moms as more of a fierce mama bear—protective, powerful, fearless, and, yes, loving. Our tagline "One Tough Mother," which we've used for T-shirts and fake tattoos, perfectly captured what I was going for.

Our goal is to use our "momness" to both appeal to and empower our target audience: women who turn into badasses when the safety of the kids in their community is threatened. As a volunteer in Little Rock, Arkansas, said, "The 'One Tough Mother' icon resonates with me (enough so that I got it tattooed on my hip), not only because the work we do is tough, but also because it helps me feel strong about my gun violence survivor story."

That's why it's important to think not just about the demographics of the people you want to attract—how old they are, whether or not they have kids, where they live, how educated they are, etc. You also want to give good thought to the psychographics of the people you're seeking to reach—what motivates them, how they think, what they care about, how they like to work. You need to put thought into both of these categories in order to craft a brand that draws your ideal people in like bees to honey.

It's also important to remember that one brand won't appeal to everyone, and that's okay. In fact, trying to be everything to everyone can dilute your brand and weaken its ability to attract people. Even to this day, some people ask why I didn't name the organization "Parents Demand Action," or something less mom-specific, but I fought for what I knew was a missing movement for an untapped audience within the gun violence prevention space. I know that it can feel risky to choose a narrow target audience, but a well-honed message will attract people who are outside that target too, and you can welcome them in with open arms.

In our membership, which we refer to internally as "mothers and others," we are proud to welcome women who don't have kids, men, and nonbinary individuals, and they are joining Moms Demand Action in droves and proudly touting our brand, regardless of whether or not they personally claim the title "mom." (Although we love the fact that many of our male members have embraced the slogan "Man Enough to Be a Mom" and proudly wear T-shirts and pins emblazoned with those words.) That's because we've communicated the value and importance of letting moms and women, the backbone of our brand, lead.

Branding Principle No. 2: Create a Compelling Look and Feel

Any sophisticated brand needs a carefully designed identity, which includes your logo, colors, and fonts. Your identity is what makes you recognizable to supporters, so it's important that it be professional, unique, and a clear representation of your purpose.

In the early days, one of our volunteers created a logo for One Million Moms for Gun Control. For the reasons I described earlier, that name was short-lived. But thankfully, at the same time we decided to change our name, I got a call from Patrick Scissons, an executive at an ad firm in Toronto, Canada, who would make the rebranding process much easier.

Patrick had read about us in the news and wanted to offer his team's pro bono support to help build our brand. It was another bit of serendipity—and, dare I say, a result of our branding efforts! Even though we were still working the kinks out and building the plane as we were flying, the fact that we clearly identified as moms helped draw to us just the right people who could help. Over the next several weeks, Patrick's team helped us hone our name and create a style guide of colors, fonts, and even graphics to use on all of our platforms. And our volunteer team of legal experts helped us be sure we owned our new name and logo.

One of the graphics we developed in those early days really captures the strength of moms that we wanted to convey: it features a female version of Superman (or Supermom) unbuttoning a very traditional button-down shirt to reveal her Moms Demand Action T-shirt, our own version of an action hero costume.

Colors are an important part of your look and feel too. Scientific studies have shown that colors alter people's emotions and that women are more sensitive to bright colors than men. Over the years we've focused on two color schemes to represent our brand: orange and red. We wear orange to represent the gun violence prevention movement. The color was chosen after the 2013 shooting death of Hadiya Pendleton, a Chicago teen who was an accidental casualty of group violence. Hadiya was shot and killed in a South Side park where she and friends had taken shelter from the rain—just one week after performing at President Obama's second inaugural parade. In her honor, Hadiya's friends wore orange because it's the color hunters wear in the woods to protect themselves and others from gunfire.

We wear red to represent the advocacy of Moms Demand Action. Not only is red stimulating to the eye, it's linked to some of the most famous brands in the world, such as Coca-Cola and H&M. Early on we found that red T-shirts popped and sold better online than less vibrant colors like blue or gray. Since then, red has become Moms Demand Action's signature color—an interesting juxtaposition with Donald Trump's Make America Great Again baseball hats.

As Kristi Cornett's story showed so well, our red shirts have become a recognizable calling card for Moms Demand Action.

Lawmakers and other influencers see our shirts from across a crowded room and make a beeline for our volunteers. According to a Boston volunteer, "In Massachusetts, we've replaced the motto of 'the red coats are coming' with the 'the red shirts are coming!' when we walk into the statehouse to lobby for gun-sense legislation."

A Portland volunteer says, "I went to a fundraiser for Oregon governor Kate Brown that was sponsored by a group of women physicians. Even though the event was focused on health care, the governor ran over to me when she saw my red shirt and made it a point to talk to me about gun safety legislation she hoped to pass during the next session. And then, in her opening remarks to the crowd, she called out Moms Demand Action by name and thanked us for being there." This is exactly what a brand does for you—it makes you more visible and more influential.

It also helps you claim your territory. One of the most powerful NRA lobbyists in the country, Marion Hammer in Florida, used to wear a red blazer in the statehouse, calling it her signature color. Since Moms Demand Action has become a powerful force in the Sunshine State, Hammer has given up red altogether. It's our color now.

Branding Principle No. 3: Amplify Your Influence with Photographs

A big reason why our brand became so recognizable so quickly is because we adopted the motto "Pictures or it didn't happen." Sure, it's important to show up. But if you show up and take a picture—and then share it on social media and in the media—people who weren't even at the event will know you were there.

In 2018, I was happy to be interviewed by Jason Kander, former secretary of state for Missouri, on his *Majority 54* podcast. And I was especially thrilled when he told me: "When I ran for the US Senate, I was at tons of events where women in St. Louis, Kansas City, and all over the state kept showing up in Moms Demand Action shirts. And it was clear that they weren't just showing up—they were clearly very thoughtful about organizing. They would ask me, 'Will you take a picture with me in my T-shirt?' I was so impressed by that branding know-how."

So yes, it matters that we show up. But it also matters that we make sure we're seen when we show up. At every meeting, whether it's for new members or with lawmakers, we take a picture together. We even took a photo with Ted Cruz, the pro-NRA and anti–gun safety Texas senator, when we showed up at one of his town hall meetings to ask questions about gun violence. Much to my surprise, and as evidence of how recognizable and desirable our brand has become, Cruz's people then included that photo in one of his campaign ads! It gave us the perfect opportunity to point out his hypocrisy and to make it clear that the only gun-sense candidate in the race was his opponent.

Branding Principle No. 4: Create a Clear Message

Brands aren't just visual; they're verbal. The messaging you create to communicate your values and beliefs should become the foundation for all of your communications, including your website, press releases, and social media posts. Clear and compelling messaging encourages a brand's target audience to care and, ultimately, to act.

The first message most people encounter from us is our name. "Moms demand action" was a phrase we'd been chanting at marches and rallies across the country right after the Sandy Hook school shooting, and "gun sense" was something Patrick's team came up with to describe our support for commonsense gun laws. We wanted to be *for* something, not against something. And, as moms and women, we wanted to be clear that we were *demanding* change, not just politely asking for it.

As one of our Tallahassee, Florida, volunteers describes our name: "The brand to me says POWER. I love all of the words. . . . 'Moms' conveys love, 'Demand' means immediacy (we ain't askin'), and 'Action' shows we're results oriented."

Although we absolutely wanted to be direct in our messaging, we also needed to be very careful to communicate a moderate, nonpartisan, and strategic stance given how fraught it can be to talk about guns in America. We needed our volunteers to have the tools they needed to organize and speak out as powerfully and persuasively as possible on both sides of the aisle.

As a volunteer from Champaign-Urbana, Illinois, said: "Moms Demand Action's messaging is very clear and inclusive of people who don't consider themselves to be for 'gun control.' Once they realize we don't oppose Second Amendment rights, it makes it easier for people—especially those in red states—to join. Many of my friends and family joined our movement after seeing how moderate our messaging is."

This need to be moderate is also why we rely so heavily on data (something the next chapter dives into)—we want our brand to convey the fact that we prioritize evidence over anecdote, and that we know our stuff. After all, our supporters and volunteers need facts to dismantle the tropes of gun lobbyists and extremists.

Once you determine your messaging, you have to make it available so that your supporters can stay true to it. We house our main talking points on our organization's websites, we have an internal clearinghouse where we keep messaging related to a wide variety of topics, and our primary messages are baked into every social media post. We're constantly saying the same five or so things over and over again—such as "This isn't a partisan issue, it's a matter of life or death" and "We all want to keep our families safe"—to ensure that what we're saying sinks in. Obviously, we have to come up with responses to new gun-related issues as they arise, but day to day, we're talking about our priorities as an organization and staying focused on specific calls to action.

Branding Principle No. 5: Connect Emotionally

There are two ways to connect with a target audience through a brand: rationally and emotionally. Our use of data helps us appeal to people's rational minds. But we can't stop there. We also need to connect to people's hearts, because it's that kind of connection that encourages loyalty and inspires action. That's why you see rational brands use ads that pull on

heartstrings to sell insurance and cars; it helps them build a strong emotional connection with their target audience. When that happens, your audience can become incredibly effective brand ambassadors—think of how devoted users of Apple products became after the company's 1997 series of ads that featured widely regarded visionaries with the tagline "Think Different." Launched even before the release of the iMac (the iPhone wouldn't debut for another ten years), those ads helped save Apple from becoming just another blip on the timeline of tech history, because they spoke to its ideal customers' self-identification as creative nonconformists. Those customers bonded with the brand, and then proselytized the company's products to their friends and families. And there is no better marketing than an enthusiastic and heartfelt recommendation from someone you know and trust.

Speaking directly to our audience's core values is exactly how Moms Demand Action's volunteer base grew so quickly—because our branding helps moms develop an emotional connection to us, and then they invite the people in their lives to come along for the ride too. Our name, our taglines, and our mission empower moms to embrace their strength and take on seemingly impossible tasks. They love us for that, and as a result, they also love the visual representations of our brand.

Kristi, from the Tennessee chapter, describes it this way: "I always feel like my red shirt makes me more comfortable. It's like a security blanket. When I'm wearing it, I feel stronger and more confident. It reminds me that I'm fighting for a greater cause."

Some volunteers say their red shirts make them feel invincible: as one Texas volunteer said, "When I put on my red shirt,

it feels like a power suit or a superhero cape. My mood changes instantly."

Now *those* are some very powerful emotions that are inspired by something as simple as a red T-shirt! And that is branding at its best.

Branding Principle No. 6: Balance Consistency and Flexibility

Consistency is an undeniably important component of creating a successful brand. Rogue fonts, low-res logos, or messaging and graphics that are out of context with the brand's purpose can make an organization seem confusing and disorganized. Imagine if Nike forgot to include its swoosh in an ad. Or if NBC sometimes used a Comic Sans font. Or if 3M occasionally spelled its name ThreeM. It would be confusing at best, off-putting at worst.

At the same time, any brand that wants to last longer than the ever-quickening news cycle has to be flexible and creative to stay true to the wants and needs of its audience.

At Moms Demand Action, we have tried to balance this need for discipline and consistency while also empowering our volunteers—our brand ambassadors—to feel ownership of this organization that is as successful as it is only because of their hard work and passion.

I knew from my corporate days how key it was to have brand consistency; what I didn't anticipate was how challenging it could be to do that when building an organization that was run by volunteers from all over the country.

In the early days of Moms Demand Action, I noticed that some of our state pages were posting stories and graphics about issues that were not related to gun safety. For example, after the US Supreme Court weighed in in favor of marriage equality, some of our state Facebook pages celebrated the win. As the mom of a gay daughter, I was thrilled with the decision, but it wasn't in our wheelhouse. Moms Demand Action focuses exclusively on gun issues. That's why we created a set of guidelines to give volunteers to hold on to as the organization's "style bible."

Our style bible establishes basic parameters of what is and isn't in line with the Moms Demand Action brand. As long as you stay within those white lines on the road, any variation is fair game.

However, in other areas, we've been more flexible. Right from the beginning, our volunteers wanted to wear their Moms Demand Action shirts everywhere—not just to gun violence prevention–related events, but to concerts, plays, their kids' sports games, and even when they volunteered for other causes. At first we were concerned about our brand being associated with anything and everything that our volunteers were interested in, but then we realized the amazing value of having tens of thousands of brand ambassadors keeping our name and cause in the public eye.

We've seen our volunteers wearing Moms Demand Action shirts at vigils and memorials; while working at food banks in Ferguson, Missouri; and in Houston while cleaning up in the wake of Hurricane Harvey. My personal favorite was the man from Austin, Texas, who wore a Moms Demand Action shirt when he was interviewed on the local TV news after he adopted a thirty-five-pound orange tabby cat named Symba from a local animal shelter.

Over the years, our volunteers have worked within our style bible (and sometimes pushed it to the limit) to put their own flair on Moms Demand Action buttons and T-shirts. Some people in the political world call this "rogue chum": *chum* refers to traditional campaign giveaway items such as T-shirts, lawn signs, and bumper stickers; *rogue* means the supporters are so passionate that they've started making the chum themselves.

First, Texas volunteers started to bling their buttons. If you've ever been to a homecoming event in Texas, you've seen the supersized, bedazzled mum corsages the girls wear to the game. Texas volunteers basically did that to their pins. When volunteers in other states started seeing photos of Texas volunteers wearing Moms Demand Action pins the size of pumpkins covered in ribbons and rhinestones, the horse was out of the barn.

Volunteers in other states started adding location-specific details to their Moms Demand Action chum: for instance, California's pin features a bear, to reflect the bear on the state's flag but also to symbolize our synergies with that state's abundance of mama bears. Other pins feature crabs or cowboy boots or corn.

And volunteers are customizing more than pins. The variations of Moms Demand Action fashion are endless. Volunteers have cropped their T-shirts, turned them into dresses, and worn them *Flashdance*-style with the necks torn out. I've seen handmade shirts that demand action on behalf of kids, grandparents, and even dogs. Some volunteers have decked out their fingernails with logos for Moms Demand Action or our Gun Sense Voter check mark. In upstate New York, volunteers had aprons made with stats about gun violence printed on them. And at some events, our volunteers even wear capes with our logo on the back.

Perhaps the most impressive way volunteers have personalized elements of our brand is to have them tattooed on their skin, whether it's the Moms Demand Action logo or our "One Tough Mother" tagline. We started off by making fake tattoos, but many, many moms have made them permanent.

I admit that, at first, my inner PR executive struggled with these variations on our carefully constructed branding elements. But it didn't take me long to realize that there really is no greater honor than for a volunteer to want to merge a piece of herself—whether it's her love of bedazzling or her actual flesh—with Moms Demand Action. We are only as successful as our volunteers are passionate, and all our rogue chum shows me that we aren't an army of anonymous volunteers; we're living, breathing, multifaceted women (and others), and no matter how big our pins are or what shape our shirts are, the better you can see us as both individuals and a united force, the more our presence will move you.

A Brand Makes You a Beacon

One of the most important things a dynamic brand does for you is make you more visible, and then you can draw more attention to your cause. And attention certainly comes in handy—as evidenced by Kristi's story of the packed room of red shirts swaying the vote at the Tennessee statehouse. Even when the stakes aren't so high, having a visible brand does some heavy lifting for you. As an Arkansas volunteer said, "When I've worn my Moms Demand Action shirt around town, I've even recruited new volunteers while standing in the grocery line!"

We also stand out at rallies and parades thanks to volunteers carrying huge, Styrofoam letters that spell out the word *MOMS*. A Missouri volunteer with a background in theater design first made the huge letters to use in the Women's March in St. Louis just after the 2016 election. "I just knew the Women's March would be big," she said, "and with my background in theater I knew we'd need something splashy to stand out. Which we did—the photographers LOVED it." Now volunteers in almost every state have their own set of these letters with their own personalization for their state (usually inside the *O* of Moms).

But in some cases, becoming more visible can also make you the target of negative reactions. One Dallas volunteer wrote on Twitter, "Wore my Moms Demand Action shirt to the bank today. Got called a 'fucking cunt libtard.' I'll wear that badge with pride." A Las Vegas volunteer who was wearing her shirt on the way to a monthly Moms Demand Action meeting stopped by a grocery store to pick up snacks. After a man stopped her in the parking lot to insult and threaten her, she tweeted this: "To the angry man at the grocery store who just called me a f**king fat c*nt and said if he had a gun right now he'd shoot me, and that I shouldn't walk to my car alone . . . you didn't break me! I'm at my @MomsDemand meeting right now and you can bet your ass we will #KeepGoing!"

Our shirts can even cause controversy when we show up in statehouses. For example, a Florida House sergeant of arms told our volunteers they couldn't wear their Moms Demand Action shirts during a gun bill hearing and that they'd have to take them off or turn them inside out. After sitting in the hearing with their shirts on inside out for a while, one of our volunteers marched up to the sergeant of arms office and demanded to be shown where it was written that our shirts weren't allowed. There was no policy to be found, and the media declared a victory for Moms Demand Action volunteers.

The good news is that for all the negative attention they can attract, our red shirts can also help you see how many allies you have in the immediate vicinity with just one glance. As one Tallahassee volunteer put it, "When I see other people wearing our shirts around town, I am deliriously happy and feel like we are united as superheroes." Another volunteer said: "When we go to rallies and we have children in strollers, we're often surrounded by men who are open carrying AR-15s, semiautomatic rifles, or Glocks. That can be terrifying. By having women show up en masse in their T-shirts, we both feel safer and look stronger as a sea of red."

As a result of our volunteers' passion and their seeming omnipresence, we're recognized everywhere we go. During the 2016 elections, politicians would look for the block of red-shirted Moms Demand Action volunteers in the audience and give us a shout-out. At one point, Bill Clinton tapped Hillary on the shoulder and, while pointing at us, said, "There are your moms!" I couldn't believe that something I started in my kitchen was being recognized by a past and (I had hoped) future president.

Being more visible promotes more than just your cause, of course: it also gives you, personally, a chance to be seen and reckoned with in a way that you may not even know you've been missing. It's a common—if often unspoken—complaint of motherhood that it can make you feel invisible. In many ways, your identity as an individual can slip away when you become a mom, whether that occurs on the surface—what mom hasn't let go of at least some of her previous interests and hobbies after having kids?—or on a deeper level—so many of us have also felt our sense of self undergo a wobble if not a major shift after the transition to motherhood. But we moms wield most of the unseen power in this country. It is absolutely right that we display that power visually, too. And when you put on the red shirt, you're impossible to ignore.

Our shirts make us recognizable to each other, too. From signs in windows to stickers on cars to our T-shirts and tattoos, when you belong to one of the largest grassroots movements in the country, you have friends everywhere.

Once during an Indiana volunteer's vacation to Asheville, North Carolina, she spotted a Moms Demand Action sign in the window of another car. She left a note on the car that said: "I am visiting from Hamilton County, Indiana. It was in our county that the Noblesville school shooting happened. I was so excited to see your sign that I wanted to take a moment to thank you for your activism! Together we will make a difference in the effort to end gun violence. I haven't met you, but together we will #KeepGoing."

And that's the final piece of what a savvy brand can do for you—lend you longevity. Because, goodness knows, the need for advocacy work isn't going to end any time soon. Any tool that helps you gather influence and build momentum is simply too vital to overlook.

8

Know Your Numbers

It's obvious that a huge piece of the Moms Demand Action effectiveness is due to the dedication of our volunteers. We make calls and send emails and knock on doors. We show up at meetings and hearings. We stay as long as we need to stay. And we keep doing it for as long as it takes to ultimately score a win. Yet, as powerful as our presence and our voices are to effect change, there's another crucial piece of the activism puzzle that has a power all its own: data.

Our three-year fight to pass a law that disarms domestic abusers in Rhode Island is a perfect example of how having trustworthy data can tip the scales in what might otherwise feel like an unwinnable fight.

According to federal law, domestic abusers subject to final restraining orders are prohibited from possessing guns, and whether they are required to surrender guns they already own is up to a judge's discretion. Under Rhode Island law, they were not necessarily prohibited from possessing guns. Nor were they always required to surrender their guns once they were convicted. Rhode Island state representative Teresa

Tanzi had worked as an advocate at a domestic abuse resource center and knew that this discrepancy between federal law and state law was costing Rhode Island women their lives. So in 2014 she introduced a bill to resolve it. The Rhode Island Moms Demand Action chapter leader, Jennifer Boylan, pledged to help.

Teresa and Jennifer assumed that the bill would be well received: a lawmaker would introduce it, other lawmakers would agree that the bill was a good idea, it would come to a vote, and it would pass. Easy.

"Looking back now," Jennifer says, "I had a very *Schoolhouse Rock*–inspired impression of how things would go. I thought it was a no-brainer. Who wouldn't want to make sure that domestic abusers didn't get to buy guns or keep the ones they already had?"

Little did she know that she, Teresa, and the Rhode Island volunteers were in for a three-year battle that would include harassment from gun extremists, thousands of hours of activism, and lots of losing forward.

That first year, the 2013–2014 legislative session, the bill failed to advance out of committee. Despite the testimony of domestic abuse survivors and plenty of calls to lawmakers from Moms Demand Action volunteers, Teresa recalls the House and committee leadership—all of them men—saying to her: "Teresa, you just don't understand. These women are only trying to game the system and get a leg up in their divorce. Judges give restraining orders out like candy. There's no real danger here. We can't take people's guns away." This is a classic example of mansplaining for you—telling a woman, who has direct experience with a problem, that there is nothing to worry her pretty little head about. When in reality, making a false claim would expose these vulnerable women to potential felony perjury charges that carry a potential jail time of up to twenty years! These men in power were demonstrating that they did not believe women or hear their cries for help, belying a fundamental and institutional distrust of women, which only drove Teresa and Jennifer to keep going.

In the 2014–2015 legislative session these two leaders upped their game and got more survivors to testify by working in coalition with Sisters Overcoming Abusive Relationships and the Rhode Island Commission Against Domestic Violence, two local organizations dedicated to supporting victims of domestic violence.

They also decided to gather some key data to help cut through the anecdotes and objectively show the scope of the problem. Teresa and Jennifer reached out to researchers at Everytown for Gun Safety, who pored over nearly twenty-two hundred court documents to analyze every case that had resulted in a restraining order in the previous two years. They found that in the 1,609 cases where Rhode Islanders were subjected to domestic violence protective orders, judges required that they surrender their guns in only about 5 percent of them—even when the person requesting the restraining order mentioned that their alleged abuser had a firearm and/or had threatened to shoot them. Everytown researchers wrote up their findings in a report that they shared with the *Providence Journal,* which, in June 2015, published a preview of those findings on the front page with the headline "In R.I. Domestic Violence Cases, Suspects Often Keep Guns."[1]

That article raised awareness of the issue and provided an important counterweight to the anecdotal narrative that women sought restraining orders just to give themselves more leverage in a divorce battle. "The report and its media coverage gave me and the Moms Demand Action volunteers something we could walk up to a legislator with and say, 'This is the reality in Rhode Island; we have to do something about this,'" Teresa recalls.

As important and eye-opening as the report was, it didn't change the fate of the bill—yet. The 2015 bill also never made it to a floor vote, because the way things work in the Rhode Island House is that nothing is brought for a vote without the approval of the speaker of the House. During this period, the speaker was Nicholas Mattiello, an A-rated NRA Democrat (in this case, getting an A is not a good thing) who was staunchly opposed to the bill. Teresa did everything she could to get Mattiello to bring the bill to a vote, including camping out outside his office for hours despite being told by his staff that he wasn't available, but he was not swayed. Luckily, neither were Teresa and Jennifer.

In the winter of 2016, Jennifer heard an interview with Speaker Mattiello on the radio. In this radio appearance, the interviewer put the speaker on the spot and asked his opinion of the movement to disarm domestic abusers. Mattiello said the bill should be moved forward to a vote. Jennifer saw this admission as an opportunity to hold the speaker accountable for what he'd said. It wasn't enough for him to suggest support of popular legislation that polled well with women; he needed to act. So Jennifer decide to up the pressure.

First, she started organizing weekly lobby days at the statehouse, where Moms Demand Action volunteers showed up in their red T-shirts to meet with lawmakers and generally make their presence known. Soon the Rhode Island team had forged new bonds with supportive legislators, especially woman lawmakers in the House and Senate who badly wanted to pass the bill. They were building a coalition of like-minded legislators; the problem was swaying those who were opposed to the idea. And that's where Everytown's report came in, again.

During the 2015–2016 legislative session, Teresa recalls that she requested so many copies of the report that the Everytown staff ran out and volunteers had to make several last-minute trips to the copy shop before each hearing. "I needed the legislators to have the facts in front of them so that they couldn't say they weren't aware of the problem, so I brought them to every hearing and gave them to everyone in attendance, including those in the audience and the elected officials."

Together, the increased presence at the statehouse and the Everytown report began to tip the scales, although not as quickly as you might expect.

It took almost a year to sway Mattiello to introduce the bill for a vote on the floor of the House in the spring of 2016. After a lengthy debate that kept Moms Demand Action volunteers at the statehouse late into the night—Teresa herself responded to a grilling from committee members for an hour and a half, a state record—the House passed the bill. It seemed like our Rhode Island volunteers would finally score a win! But when the bill moved to the Senate, a budget disagreement abruptly ended the session before the bill could come to a vote. The bill was in limbo until a special session reconvened.

Finally, in September 2017, after three years of losing forward, both houses of the Rhode Island legislature passed the Protect Rhode Island Families Act. A few weeks later, the

Moms Demand Action volunteers who'd advocated for the bill stood with Teresa and watched as a woman and fellow mom, Governor Gina Raimondo, signed the bill into law.

"The Protect Rhode Island Families Act would not have passed without the efforts of the Moms Demand Action volunteers—that's a fact," Raimondo says. "The legislative leadership had no appetite to pass this bill until Moms volunteers showed up and protested and testified and made themselves impossible to ignore. They changed the conversation."

Governor Raimondo's support for the bill also played a big role in its ultimate passage; it was a unique opportunity to be able to partner with the governor during the process. Having her support inspired even more women to get involved, particularly when it came time to testify at various hearings. "Testifying for or against a bill can take hours; you have to put your name on a list, and you might be number 110," Raimondo explains. On the days when this bill would be heard, she made the governor's suite at the statehouse available to women bringing children—she had a large closet painted and converted into a very comfortable nursing station. She also arranged for snacks and babysitters. "We wanted women to know that we want you here, we need you here, and the process will be better if you're here," Raimondo recalls. It's a strategy that paid off: "The fact that so many women had the courage to tell their stories was unbelievably powerful. I could tell it made a difference in the lawmakers' minds." (Which is just one more reason why we need more women in office—to make the legislative process more welcoming to other women.)

While what Teresa calls "the unceasing drumbeat of the women in red shirts showing up day in and day out" and the bravery of so many women sharing their harrowing stories of domestic abuse were both major factors in that victory, I fully believe that it was the data that ultimately began to sway the speaker's mind and pushed the bill to a vote. Just as you should never underestimate the power of a group of passionate women, you should also never neglect to know what the numbers are and when to wield them. After all, any argument is just an opinion until you have the research and data to back it up. Once you have reliable numbers, that opinion transforms into fact.

How We Use Data to Set Our Priorities

At Moms Demand Action, we have passion in spades—I know you have it too, or you wouldn't be wading into a long chapter about the power of numbers! But passion can quickly dissipate when you're dealing with a problem that has more contributing factors than an octopus has legs. Just when you think you've made progress on one leg, another starts waving around, diverting your attention and scattering your efforts. Data help you see the whole octopus so you can decide which one or two (or, in our case, three) legs you want to prioritize.

Moms Demand Action bases its policy platform on data-driven research that focuses on one specific outcome: what will save the most lives. On the basis of the available evidence, we know three policies that have proved to be extremely effective at reducing gun deaths: closing loopholes in the background check system, keeping guns out of the hands of known domestic abusers, and passing red flag laws (which allow for the temporary removal of guns from individuals who have shown patterns of violence). These policies are effective, and they enjoy bipartisan support, which is why we've made them our top policy priorities.

Certainly, other policies are important to our volunteers, including restrictions on assault weapons. Moms Demand Action certainly supports limitations on assault weapons to restrict access to firearms like the ones used in the Parkland, Las Vegas, and Pulse shootings, as well as most of the deadliest mass shootings in recent history. We've been proud to stand alongside local gun safety advocates who've helped pass commonsense restrictions on assault weapons in places like Boulder, Colorado, after the Parkland shooting and Washington State in the 2018 midterms. But while assault-style weapons certainly put the "mass" in mass shooting, the data show that rifles are responsible for only about 3 percent of gun-related homicides; most firearms deaths and injuries are caused by handguns. That's why we keep our focus on our top priorities; if we spread ourselves too thin, our focus and our progress on the biggest sources of lost lives will be diluted. In advocacy, just as in parenting, you have to pick your battles.

It's not just about figuring out what to prioritize though. Being armed with data helps in so many ways; research is like the Swiss Army knife of activism:

- **It helps sway lawmakers' minds.** When you can point to numbers from verifiable sources (more on how to determine whether your data are credible on page 183), you present a much more compelling case. Numbers offer lawmakers protection—if they can point to a statistic, they can explain their votes to someone who may not be happy with their decision. The reports we've put together on domestic violence, for example, have helped turn lawmakers from stallers into supporters (Speaker Mattiello is a case in point, but he's by no means the only one).

- **It inspires volunteers.** When you know that what you're working on will save lives or change the culture—and you can see demonstrable proof of that power in the numbers—you will be motivated to keep doing the heavy lifting of organizing and advocating. Numbers give you inspiration; they also give you gratification when you see them change over time.
- **It gives you credibility.** Let's face it: as women, we're often dismissed as being too emotional or even hysterical. Knowing the facts makes you more credible, whether you're talking with a relative at the Thanksgiving table or hashing it out with someone on social media. The facts take ammunition away from your detractors and can provide a counterbalance to narratives that may or may not be wholly accurate.

Finding the Facts on Gun Violence—It's Complicated

When it comes to gun violence, it can be challenging to find reliable data. This is thanks in large part to the efforts of NRA lobbyists, who worked in tandem with members of Congress beholden to them to essentially muzzle the Centers for Disease Control and Prevention (CDC)—the arm of the government that once funded research on the prevalence and impact of gun violence in the United States.

This push began shortly after a CDC-funded study was published in *The New England Journal of Medicine* in 1993. In it, researchers showed that instead of protecting you, having a gun at home made you more vulnerable to homicide[2]—a blow to the whole "guns make people safer" messaging of the NRA. And so the NRA and its minions went all out to shut down any more research that would likely be equally damning.

With the NRA's advocacy, the US House of Representatives passed an amendment to the federal budget in 1996 that essentially gutted the CDC's ability to research gun violence. The amendment removed $2.6 million from the CDC's budget—precisely the amount it had previously spent on gun violence research[3]—and contained the caveat that "none of the funds made available for injury prevention and control at the Centers for Disease Control and Prevention may be used to advocate or promote gun control."[4]

Even though these efforts targeted only the federal government, they made gun violence a third rail for academic researchers too. Nobody wanted to touch it. Nobody, that is, except John R. Lott Jr., an economist and a prolific—if flawed—researcher who is the go-to guy when gun advocates want to bolster their argument that guns make people safer. Lott first came on the scene in 1997 when he was a research fellow at the University of Chicago. That year, he co-wrote a study in which he stated "concealed handguns are the most cost-effective method of reducing crime thus far analyzed by economists"[5]—a premise he expanded upon at length in his 1998 book *More Guns, Less Crime.* That factoid—which has now been thoroughly debunked by multiple other studies[6]—lodged in the brains of lawmakers. They cited it in their quest to pass looser gun laws, and Lott even testified in favor of concealed carry laws before at least five state legislatures.

As Lott became more heralded, other researchers—from institutes as storied as Stanford, Yale, Harvard, and the National Research Council, a division of the National Academy of Sciences—sought to verify his claims, and couldn't. It seems Lott overlooked a lot of data and didn't take into consideration that crime in the 1980s and 1990s spiked because of the crack epidemic, not because of laws that prohibited concealed carry.[7]

Guns aren't the only topic where Lott has released selectively sourced data that conservative politicians have taken and run with—he's also a climate change denier and has published erroneous data that suggest that undocumented immigrants and Dreamers are more likely to be convicted of serious crimes and that legal abortion leads to higher crime rates.[8]

Even though Lott's gun data have been discredited, they are still in circulation and are cited by lawmakers and lobbyists so frequently that the data refuse to die. (It doesn't hurt that Lott is a frequent guest on television news outlets.) And that's just one more reason to know your numbers—you need to be able to disprove those who will attempt to manipulate the data in order to further their own objectives.

At Moms Demand Action, we are incredibly lucky to have Everytown for Gun Safety as our primary research arm—after all, Michael Bloomberg built his fortune by gathering, analyzing, and sharing data on his Bloomberg terminals. His mantra, whether he was running the Bloomberg company or serving as the mayor of New York City, has always been, "If you can't measure it, you can't manage it"—a man after my own heart!

Since he was mayor of New York City (from 2002 to 2013), Michael has devoted resources to researching gun violence—both by hiring staff for his administration who analyze reports from police, the FBI, and other crime-prevention organizations, and by donating money to academic institutions (such as the Johns Hopkins Bloomberg School of Public Health) that then provide a safe haven for academics to conduct their own research.

One of those staff members is a hero to me: John Feinblatt was Bloomberg's policy director and is now president of Everytown. John was instrumental in getting marriage equality passed in New York—a campaign we've used as a template for our gun safety strategy—and he's been an amazing general for our grassroots army.

Under John's direction, Everytown compiles incredible amounts of data and creates clear, compelling reports on a variety of particular aspects of gun violence—the efficacy of background checks, for example, or the relationship between gun violence and women—and publishes them at the website everytownresearch.org. Any time you want to have the latest reliable facts on gun violence on hand, make this website your first stop. Our volunteers certainly do.

How to Know Which Data to Trust

As important as data are, numbers and statistics still can be slippery. As John R. Lott Jr. demonstrates, they can be manipulated to tell many stories. But he's certainly not the only one who does this. There is even a well-known scientific principle—known as *confirmation bias*—that asserts that humans have a tendency to look for evidence that supports or proves what they already suspect to be true.

And it's easier than ever to read something on social media and click a button to share it with your network—only to find out later that it was misleading or even downright untrue.

So how can you tell which information is trustworthy and which information isn't?

From the tireless and incredibly smart folks who work in the Everytown research department, I've learned to ask the following questions about any data. I hope they will help you find the numbers you can put stock in and use to guide your actions.

- **Where do the researchers work?** In general, it's a good sign when the authors of a study are affiliated with academic institutions that have a track record of conducting credible science (although this isn't a guarantee of objectivity—see the "Who funded the research?" bullet for more). If the researchers work at a think tank, does it have a specific political point of view? A key word to look for when researching a nonprofit organization where a researcher works is "nonpartisan," which generally indicates that the group is trying to be as objective as possible and not put forth a specific, politically determined agenda.
- **Where are the data published?** In a peer-reviewed publication, where other experts in the field read the study and challenge any conclusions they find questionable before it is published? Or on an organization's or individual's own website? Generally, the more objective eyes that have reviewed the findings before they are published, the better.

- **Who funded the research?** As helpful as it can be to see which academic institutions are involved in the study, it's no guarantee that this means the research is objective. You want to know, for example, whether research on the safety of a pesticide is being funded by the company that manufactures that pesticide (as was the case with Monsanto and esteemed horticultural academics at the University of Florida), or whether a study that finds that exercise is a more effective weight-loss tool than cutting calories was funded by a soda manufacturer (as Coca-Cola was found to do in 2015).[9] When money is involved, motives can get convoluted, and results can get tainted. If someone shares a study with you, do a little digging to see whether you can find who paid for the study. It could be as simple as looking for the phrase "This research was funded by ____."
- **How is the subject defined?** In order for a number to be meaningful, you need to know how the researchers defined the question they were seeking to answer. For example, to an economist, the definition of "the economy" is generally "gross domestic product." A layperson, on the other hand, thinks of "the economy" as "jobs and wages"—two totally different things. So if an article or a politician says, "The economy has never been stronger," you have to determine how "the economy" is defined in order to be able to compare today's "economy" to the same "economy" from a year ago.

  Likewise, to some people, a "school shooting" means any firing of a gun that happens on school grounds. To others, it means only an incident like Sandy Hook or Parkland, where a shooter makes a targeted attack on students. Understanding the definition of the subject that's being quantified is crucial to being able to understand and trust data.

Dispelling the Biggest Myths About Gun Violence with Data

One important way we can help change America's deadly gun culture is by pulling back the curtain on the common misperceptions about guns, gun violence, and gun safety laws in this country. What follows are the most pervasive myths swirling around in people's minds and on their social media feeds, as well as the numbers that dispel the myths. May these help you start having conversations with friends, family members, neighbors, and legislators on both sides of the issue.

MYTH: More Guns Mean Less Crime

Fact: There are no two ways about it: the United States is experiencing a gun violence crisis, and it is happening despite the fact that 393 million guns are in circulation among American civilians.[10] If more guns and fewer gun laws made us safer, we would be the safest country in the world. But we aren't—not by a long shot! (Pardon the pun.) According to a 2016 study published in *The American Journal of Medicine*, our gun homicide rate is 25.2 times higher than that in other high-income countries; for fifteen- to twenty-four-year-olds, that rate is 49 times higher.[11]

We're not just killing each other with guns more than in other countries; we're also killing ourselves: our gun-related suicide rate is eight times higher than that in other high-income countries.

If you really want to get depressed, outraged, and, I hope, engaged, you should also know that the overall death rate for women in the United States from firearms is 90 percent higher than that in other high-income countries; for infants to children aged fourteen, it's 91 percent higher, and for youth aged fifteen to twenty-four, it's 92 percent higher. All in all, 82 percent of the gun deaths that occur in a single year in all developed countries worldwide happen in the United States.

MYTH: Organizations like Moms Demand Action Just Want to Take Your Guns Away

Fact: Moms Demand Action is anti–*gun violence*, not anti-*gun*. Many of our members own guns. We don't want to eradicate the Second Amendment. What we want are legally mandated responsibilities to go along with the right of gun ownership, and we want commonsense gun laws that keep guns out of the hands of people who have a demonstrable risk of being dangerous.

This myth and the fear behind it is a tactic used by the NRA leadership to try to boil the issue down into black-or-white terms; according to them, either everyone gets a gun with no restrictions whatsoever or nobody does. Given that choice, many people will choose gun ownership. But this is a false choice. And it isn't a choice that even people who are members of the NRA want: in 2012, Frank Luntz, a conservative Republican pollster, polled NRA members and found that more than 74 percent of them support requiring criminal background checks for anyone purchasing a gun, and 87 percent believe that the Second Amendment needs to be balanced by measures to keep guns out of the hands of criminals.[12]

MYTH: Strong Gun Laws Don't Work; Look at Chicago

Fact: People often place the strong gun laws in Illinois next to the continued gun violence in Chicago as proof that such laws don't work. But the fact is, Indiana, where I used to live, has incredibly weak gun laws and is right next to Chicago. All someone has to do is get in their car in Chicago, drive twenty minutes to a gun show in Indiana, load up their car with dozens of guns with no background check required, turn right around, and sell those guns to whomever back in Chicago.

This is not just conjecture: a report done by the City of Chicago found that 60 percent of guns recovered from crime scenes in that city came from out of state, primarily Indiana.[13] Unless we build a wall around Chicago, guns are going to cross state borders as easily as cars do. That is why, even though it's important to advocate for better state laws to regulate guns, we also fight for federal laws. Because the way things stand right now, states with strong gun laws that are next to states with weak gun laws are much more vulnerable.

MYTH: Criminals Will Always Find a Way to Get Their Hands on a Gun

Fact: Background checks stop gun sales to criminals every single day. In fact, between 1998 (when the system was implemented by the FBI) and 2016, background check laws blocked more than three million gun sales to people who could not legally own guns.[14]

You can also see evidence of how well background checks work to save lives by looking at two states in particular: Connecticut and Missouri. In 1995, Connecticut passed a law that required gun buyers to get permits—a process that required a background check. The passage of that law is associated with a 15 percent decrease in suicides and a whopping 40 percent decline in gun homicides—and these results are cumulative over the first ten years of the law's existence.[15]

On the other hand, Missouri repealed its background check requirement for handgun purchases in 2007 (a law that had been in place since 1921). Researchers at Johns Hopkins University analyzed death certificate data between 2007 and 2010 and found that the state had seen a 25 percent increase in gun homicides. They also analyzed crime-reporting data from the police through 2012 and saw a 16 percent rise in the murder rate. To make matters worse, the state saw a 200 percent increase in the percentage of guns that had an unusually short time between retail sale and recovery by police—an indicator of gun trafficking that implies that significantly more guns were sold to more people who otherwise would have been prohibited from buying a gun via a licensed sale.[16]

For these reasons, supporting state and federal gun laws that close background check loopholes is our number one priority.

MYTH: You Already Have to Get a Background Check; What More Do You Need?

Fact: The federal law requiring a criminal background check before a gun sale can be completed covers only licensed sales. So if you buy a gun at Walmart, yes, you do have to get a background check in every state; but millions of guns are sold every year without a background check through unlicensed sales, through online sales, and at gun shows. You can buy a gun at a garage sale, online, or at a gun show from an unlicensed gun seller with no background check in thirty-one states.

As of this writing, twenty states have laws on their books to close the loopholes in the federal laws requiring a background check. Much more work needs to be done to ensure that more gun sales follow the background check process.

MYTH: The Only Thing That Will Stop a Bad Guy with a Gun Is a Good Guy with a Gun

Fact: The truth is that in a high-stakes situation, it is very difficult even for trained professionals to hit their intended targets with a bullet. Even New York City police officers have only a 34 percent accuracy rate.[17] Also, armed security professionals were present during the mass shootings at the schools in Santa Fe, Texas; Parkland, Florida; and Columbine, Colorado; and at Virginia Tech—and they couldn't stop the carnage.

The NRA likes to tell a story about a man who allegedly saved the day during a 2017 mass shooting at a church in Sutherland Springs, Texas, where twenty-five people and an unborn baby were shot and killed and twenty others were wounded. On this November Sunday, a man who was having an argument with his mother-in-law suited up in tactical gear and a bulletproof vest and went to the church where she attended services (although she was not present that day) and launched his attack on the congregants. A local resident with a rifle confronted the shooter *after* he exited the church and pursued him by car. Ultimately, the shooter was shot twice by this "good guy with a gun"—in the thigh and in the torso—before the shooter shot himself in the head and died.[18]

The NRA went to town claiming this as a win for "good guys with guns." And while the man who chased the shooter did display bravery, is this what has to happen in order to be considered a "win" in America today?—a shooter has to kill twenty-six people and wound twenty others before killing himself?

Most people don't realize that it takes a ton of training and experience to be able to use a weapon with any accuracy in a stressful situation. The military requires extensive training before a soldier can carry any new weapon and requires yearly requalification on that weapon for the soldier to continue using it, since knowing how to shoot a gun and shooting it accurately are skills that can wane over time. Yet every single year the NRA is trying to strip from state laws mandatory training around gun ownership.

MYTH: Mass Shooters Target "Gun-Free Zones"

Fact: This myth is closely related to the myth that "only a good guy with a gun can stop a bad guy with a gun" by suggesting that mass shooters (a mass shooting is defined by the FBI as an incident in one geographical location where four or more people are killed by gun violence,[19] although a federal law defines it as an incident with three or more victims[20]—we use the FBI's definition) seek out places where guns are prohibited for the scenes of their crime. Evidence clearly refutes this idea.

Everytown for Gun Safety analyzed all 156 mass shootings that occurred between 2009 and 2016 and found that only 10 percent happened in a "gun-free zone." It also found that most of these instances—63 percent—occurred in private homes, which makes sense when you learn that more than half (54 percent) of mass shootings are related to domestic or family violence.[21]

MYTH: Guns Don't Kill People, People Kill People

Fact: This is exactly why we fight so hard for background checks on *people*; if people kill people, no one should be able to buy a gun without first undergoing a criminal background check. This is the most effective way to keep guns out of the hands of people who have dangerous histories. We know that commonsense laws to keep people such as domestic abusers and violent criminals from having guns save lives. And it's also why we want mandatory gun safety training for concealed permit holders. Guns in the hands of the unskilled is a recipe for disaster.

MYTH: Gun Violence Isn't a Gun Issue, It's a Mental Health Issue

Fact: The whole world struggles with mental health issues. According to the World Health Organization, the Americas have only 21 percent of worldwide cases of anxiety and 15 percent of incidences of depression,[22] but the United States alone has a gun murder rate that is twenty times higher than that in other developed countries. Blaming mental illness for gun violence is the same as blaming movies, video games, or the culture at large—it's a tactic meant solely to distract from the real issue, which is easy access to guns, even for people with a history of criminal activity, and weak gun laws. In fact, people with mental illnesses are much more likely to be *victims* of violent crime than perpetrators.[23]

However, many mass shooters have a history of mental illness—in fact, 42 percent of mass shooters "exhibited warning signs before the shooting indicating that they posed a danger to themselves or others."[24] This makes it all the more important that we pass red flag laws. Some of these laws, such as the one passed in Delaware in 2018, also allow mental health professionals to report people who are exhibiting dangerous behavior and procure a court order to have their guns removed temporarily.[25]

MYTH: Gun Ownership Is a Constitutional Right, so It Can't Be Regulated

Fact: Constitutional rights *can* be regulated—after all, you can't yell "Fire!" in a crowded theater (a regulation of the First Amendment right of free speech). Even Antonin Scalia, the stalwart conservative Supreme Court justice who died in 2016, wrote in his opinion of the landmark case *District of Columbia v. Heller*—in which the Court upheld an individual's right to have a gun, striking down the District's ban on handguns—that the Second Amendment can be regulated: "Like most rights, the Second Amendment right is not unlimited. It is not a right to keep and carry any weapon whatsoever in any manner whatsoever and for whatever purpose." Pressing his point further, Scalia continued: "The Court's opinion should not be taken to cast doubt on longstanding prohibitions on the possession of firearms by felons and the mentally ill, or laws forbidding the carrying of firearms in sensitive places such as schools and government buildings, or laws imposing conditions and qualifications on the commercial sale of arms."[26]

The NRA will talk about the right to bear arms as a God-given right, as if God wrote the Second Amendment. But it is instead a human-given right, a right given by a government, and it can be regulated by the same.

MYTH: Guns Make Women Safer

Fact: Women in the United States—which has more guns than the next twenty-five countries combined—are eleven times more likely to be killed by a gun than women in other high-income countries.[27] This is in large part because of the fact that if a gun is readily available during a domestic violence situation, a woman is five times more likely to be killed. In an average month, fifty American women are shot to death by an intimate partner, and nearly one million American women alive today have been shot or shot at by an intimate partner. If that's not bad enough, four and a half million American women alive today have been threatened by an intimate partner with a gun.[28]

Here's what *does* make women safer: background checks. In fact, the National Instant Criminal Background Check System (NICS) has blocked more than three hundred thousand gun sales to domestic abusers since its inception in 1998,[29] and it continues to save lives every day: one in seven unlawful gun buyers stopped by a federal background check is a domestic abuser.[30] In states that go beyond federal law and require background checks on all handgun sales, the rate of women shot to death by intimate partners is 47 percent lower.[31]

The problem is that there are too many gaps in federal and state gun laws that leave women vulnerable:

- Although background checks do stop abusers from purchasing guns every day, as I've covered, the system doesn't apply to online sales, sales at gun shows, or sales from unlicensed dealers. In a 2013 report, Mayors Against Illegal Guns found that one of every thirty people who sought to buy a gun through the website armslist.com had a criminal record that would prohibit them from purchasing a gun if they were required to pass a background check—meaning that an estimated twenty-five thousand guns were sold to criminals who had felony or domestic abuse records through this site.[32]

- There are two important categories of domestic abusers that laws typically don't cover: convicted stalkers and abusive dating partners (known as the "boyfriend loophole"). This is especially alarming considering that more American women are killed by people they are dating than by their spouses.[33]
- Many times, state law doesn't back up federal law. Without matching provisions in state law, local law enforcement has no way to enforce federal restrictions. Federal law prohibits convicted domestic abusers from possessing a gun, but it doesn't affirmatively require them to turn in any guns they already own. Neither do the laws in thirty-five states. *Thirty-five states!*[34]

We still have a long way to go on keeping women safe from guns, particularly in domestic abuse settings, which is why working to get state laws passed that provide clear, enforceable laws that do just that is so important. The good news is that since 2012, twenty-eight states plus the District of Columbia have strengthened their laws to keep guns away from domestic abusers. And we are not letting up anytime soon.

MYTH: Arming Teachers Will Make Kids Safer

Fact: The push to arm teachers isn't, at its root, about keeping kids safer. It's about selling more guns. Since Donald Trump was elected president, gun sales have plummeted because there's no boogeyman in the White House to make people afraid that their guns will be taken away from them. Known as the "Trump Slump," this downturn in gun sales is hurting gun manufacturers: American Outdoor Brands, the manufacturer of Smith & Wesson guns, announced a 32 percent decline in sales for the third quarter of 2017, and giant gun maker Remington filed bankruptcy in February 2018, saying that it had between $100 million and $500 million in debts.[35]

There are 3.6 million teachers in the United States. Arming even a fraction of them would go a long way toward replenishing the coffers of gun manufacturers and, in turn, the NRA.

The other market being targeted in this attempt to "harden" schools is kids. If every kid gets used to seeing guns in school, they'll be that much more likely to want to own a gun as an adult. Worse yet, the plan that Secretary of Education Betsy DeVos is considering as I write this book would use funds earmarked for the Student Support and Academic Enrichment program in the country's poorest schools to buy firearms and provide firearm training. Frankly, it's hard not to see this push as an attempt to further defund and destabilize public education.

On top of all this, schools could never be armed enough to prevent deadly school shootings. Remember, armed guards were present at the schools in Santa Fe, Texas; Parkland, Florida; and Columbine, Colorado; and at Virginia Tech—and they weren't able to stop the murder of innocent kids and others. And these were trained security professionals. How would teachers, who are trained to educate, be able to keep their cool in such a stressful situation and take out a shooter without injuring—or worse, killing—someone who isn't their intended target? Don't forget, even New York City police hit their intended targets less than 34 percent of the time.

MYTH: The Pockets of the NRA Are Just Too Deep; It's Pointless to Fight Against It

Fact: The NRA itself admits that it is in financial trouble. The organization ended 2017—the year of its most recent IRS tax filing at the time of this writing—with a deficit of nearly $32 million. Compare that to 2015, which it ended with a $27.8 million surplus.[36] The Trump Slump, which is hurting gun manufacturers, is in turn hurting the NRA, which is primarily funded by those manufacturers.

Part of the NRA's financial woes can be linked to the actions of New York State financial regulators, who in 2018 ruled that an insurance policy the NRA called "Carry Guard"—intended to reimburse NRA members who incurred legal costs as the result of firing a legal gun—"unlawfully provided liability insurance to gun owners for certain acts of intentional wrongdoing."[37] As a result, the NRA's insurance partners were forced to pay a $7 million fine and discontinue the policy.

In the wake of this investigation, the New York Department of Financial Services cautioned other insurance companies and financial institutions in the state to "review any relationships they have with the NRA or similar gun promotion organizations, and to take prompt actions to managing [*sic*] these risks and promote public health and safety."[38]

The NRA responded with a lawsuit against New York governor Andrew Cuomo, which complained that the organization had been subjected to a "blacklisting campaign" that had caused "tens of millions of dollars in damages" and that could soon render it "unable to exist."[39] (Coverage of this filing in the media by, among others, *Rolling Stone* prompted many people on Twitter to offer "thoughts and prayers" to the NRA—sometimes irony is just delicious.)

After Parkland, other US businesses severed their ties with the NRA: the Bank of Omaha discontinued its NRA Visa card, and dozens of companies dropped their NRA discounts, including Enterprise Rent-a-Car, Allied and North American Van Lines, Avis, Budget, Hertz, Best Western Hotels, Wyndham Hotels, Delta Air Lines, and United Airlines.[40] The tide has already begun to turn on the NRA's financial health and its credibility in the business world.

The NRA's influence in the political realm has started to wane too. The election of Donald Trump notwithstanding—it spent $30 million on his campaign—the money it spends to sway elections is having less of an effect. In 2017, the NRA spent $2 million in Virginia supporting candidates who were anti–gun regulation, and it lost every major race in that state; the governor, lieutenant governor, and attorney general who were elected all support gun-sense legislation. The Virginia House of Delegates also flipped to blue. Gun violence was one of the top three voting issues.

Virginia Moms Demand Action volunteers can take a lot of credit for these results. They made thousands of phone calls, showed up at campaign rallies, and held nearly one hundred voter canvasses.

The NRA has had diminishing returns on its investment in elections for several years: in 2012, 95 percent of the $18.3 million it spent on that election cycle was on elections its candidates lost.[41]

Looking at the Data Can Change Your Perspective

One final thing that research and data have done for me personally and for Moms Demand Action as a whole is to open our eyes to the reality of how much bigger the issue of gun violence is for people of color in the United States.

I fully admit that I was living in a bubble when I started this journey. As I've said, my biggest worry was whether my kids would make friends and do well in school. And seeing the data made me realize just how many moms were afraid every day that someone they love would be shot. I go into more detail in Chapter 9, but to give you a preview, black Americans are ten times more likely than white Americans to die by gun homicide.[42] If you are white and you have any level of fear that your kids and loved ones are vulnerable to gun violence, multiply that feeling by ten to start to get a glimpse of how much worse things are for African Americans.

When you start to become more interested in data, remember to keep an open mind and prepare to see stats that you never even thought to measure—it will change your work, and it will change you. It certainly changed our organization and helped us realize we had to build a bigger, more inclusive tent. Whether you're building an organization or getting involved in a cause, it's important that you look at the numbers through the lens of all the different populations that are affected by the issue you're seeking to improve.

9

Build a Big Tent

For me and so many of our volunteers, the Sandy Hook shooting was a defining moment in our lives—the realization that children as young as six years old could be murdered inside the sanctity of their elementary schools was more than we could bear. It made gun violence personal for us. It catalyzed us to take action, to do whatever we could to prevent such shootings from happening in the future.

For you, something else might have made you say "enough is enough," whether it was Parkland, Las Vegas, the shooting at the Tree of Life synagogue in Pittsburgh in October 2018, or a personal connection to another issue. Perhaps your catalyst hasn't happened yet. Every activist, working on any cause, has their own unique experience inspiring them to get involved. Whatever it is, that moment feels incredibly personal. But it's important to remember: even though you may feel that you have found "your" cause, it is not yours alone. Nor is your moment more important than someone else's turning point moment. To take on any issue effectively, you have to keep your ears, eyes, and heart open to the experiences of others and welcome diverse groups of people into your fold. Being inclusive and equitable—and not just exerting your will, which, well-intentioned though it may be, is limited by your own point of view—is the only way to create change that benefits everyone.

When you build a tent big enough for every population that's affected by and cares about your cause, you combine your collective power into something bigger than any one group. Inclusion, intersectionality, and diversity aren't just buzzwords—they're a key component of a successful organization's strategy.

Waking Up to Injustice

I admit that in the early days of Moms Demand Action, I was thinking of gun violence very narrowly, as primarily a problem centered around mass shootings. I had a lot of learning to do—a lot of stepping outside my own bubble and seeing how my understanding of gun violence was so starkly different from the reality that many Americans live every day.

The truth is, people have a tendency to shoot at what scares them, and sadly, in the United States, that often means shooting at people who don't fit the white, Christian, heterosexual mold, particularly black, brown, LGBTQ, Jewish, and Muslim Americans.

As a white mom of five, I've never had to tell my children to be careful around law enforcement officers because if they make any "wrong" move, those officers might shoot them. I haven't had to have "the talk" with them about how people may perceive them as a threat simply because of how they look, making them more likely to be shot. I haven't prayed every time they left the house that they'd return home safely. I haven't worried that they'd get shot in the streets of our neighborhood. Yet so many Americans face these realities every day.

I had been blind to this truth until I immersed myself in the gun violence prevention movement. Seeing the data helped open my eyes: black men are sixteen times more likely than white men to be victims of gun homicides;[1] black children and teens are fifteen times more likely to be the victims of gun homicides than white children and teens;[2] black women are twice as likely to be fatally shot by an intimate partner compared with white women.[3] And black Americans have to live with the very real possibility that any interaction with police could end disastrously: black men are three times more likely to be shot and killed by police officers than white men.[4]

One of the police shootings of an unarmed African American that stuck with me the most is the story of Stephon Clark, a black father of two who was shot in March 2018 in his own backyard by Sacramento police. Police said they thought Stephon had a gun when they confronted him. But he was holding only a cell phone—no firearm was recovered from the scene.

And it wasn't just that he was shot in his own backyard; he was shot eight times. That is shooting to kill—not seeking to de-escalate the situation or use lesser force.

Stephon's grandmother said: "They didn't have to kill him like that. They didn't have to shoot him that many times. Why didn't you shoot him in the arm? Shoot him in the leg? Send the dogs. Send the taser. Why?" Even months later, her simple question is in my mind: *Why?*

Many black and brown people believe that our country's toxic gun culture and implicit bias led to the death of Stephon Clark—and I agree. Stephon's grandmother should not have to plead with law enforcement to "only" shoot black youth in their legs or arms. Black and brown mothers should not have to live in fear that their children will be shot because they are carrying a cell phone. Something is seriously wrong when this is the reality for entire communities.

I have come to see, very clearly and without the slightest shred of doubt, that the fight for gun violence prevention is also a fight for equality. And one of the most powerful levers we at Moms Demand Action have available to us is to work on getting state stand-your-ground laws blocked, watered down, or repealed.

Taking On Stand-Your-Ground Laws

One of the most dangerous threats to people of color when it comes to gun violence are so-called stand-your-ground laws. These ordinances build on the centuries-old legal premise of self-defense—which supports the use of force to fend off an attack. One important component of self-defense has been to seek a way to de-escalate or avoid conflict or use a lesser degree of force whenever possible. The only exception to this "duty to retreat" was inside private homes, where it was legal for people to "stand their ground." Meaning, if you are attacked in your home, you don't have an obligation to try to de-escalate; you are within your rights to use deadly force first.

Florida, which has earned the nickname the "Gunshine State" for its willingness to pass laws that support unfettered gun rights, passed a law in April 2005 that made the stand-your-ground principle apply in all public places. It was the first law of its kind, and it came straight from the NRA—its lobbyist, Marion Hammer (the woman who used to wear a red blazer until we claimed that color), was instrumental in its creation and passage.[5] According to the law, Floridians don't have a "duty to retreat" anywhere. No matter where they are, if they perceive a threat, they can shoot to kill.

Stand your ground basically codifies a "shoot first, ask questions later" approach. And the result is devastating for communities of color. Research has found that in Florida, a defendant is twice as likely to be convicted in a stand-your-ground case if the victim is white than if the victim is not white.[6] Across the country, if a white shooter kills a black victim, that shooting is eleven times more likely to be ruled "justifiable"—in other words, blameless—than if the shooter is black and the victim is white.[7]

Stand your ground is the reason that George Zimmerman—who, on February 26, 2012, shot and killed Trayvon Martin, an unarmed seventeen-year-old whose only crime was walking down the street as a black person—was acquitted. The police in Sanford, Florida, where the shooting occurred, at first didn't even file charges because, in the words of the city manager, "By Florida Statute, law enforcement was PROHIBITED from making an arrest based on the facts and circumstances they had at the time."[8]

The 2005 Florida stand-your-ground law was the first in the

country, but it wasn't the last—since then, twenty-four other states have followed suit.[9] The law essentially gives regular citizens more authority to kill people than US soldiers in war zones, who are instructed to use "graduated measures of force," and law enforcement officials, who are required to seek to de-escalate confrontations first and use deadly force only as a last resort.[10]

When these laws are passed, legislators often give impassioned speeches about how the laws will reduce crime, but research has shown otherwise: A 2012 study by economists at Texas A&M University that analyzed FBI Uniform Crime Reports from 2000 to 2010 in states that had adopted stand-your-ground laws and those that hadn't found no change in the number of burglaries and robberies in the stand-your-ground states. But they *did* find a statistically significant rise in all homicides—8 percent more, which many not sound like a lot until you consider that that means six hundred additional people killed in each stand-your-ground state.[11] This law isn't about saving lives; it's about taking them—particularly black and brown lives—without any legal repercussions.

As an advocate for change in our nation's current gun climate, I, and every Moms Demand Action volunteer, have the duty to assist the effort to repeal these laws and prevent more being passed. So far, only one state, Louisiana, has repealed stand your ground. As I covered in Chapter 4, in 2017 Florida expanded stand your ground to make it easier for shooters to claim they are protecting themselves, and in 2018, a South Florida appeals court upheld the law's constitutionality despite the best efforts of our Florida chapter.

We did help to block the passage of a stand-your-ground bill in Utah in March 2018, and the two thousand signatures we delivered to Ohio lawmakers helped sway them to delay a vote on a similar bill in June of that same year. But we have so much more work to do. And we need to do it quickly, because hate crimes are on the rise.

Gun-related hate crimes in the United States occurred more than 10,300 times a year from 2006 to 2015—that's more than twenty-eight each a day.[12] More than half of them—58 percent—are motivated by racism; more than 20 percent by bias against a religion, usually Judaism or Islam; and nearly 20 percent by prejudice against sexual orientation or gender identity.[13]

Since 2015, the numbers of hate groups and hate crimes have increased; for example, anti-Muslim hate groups tripled in 2016.[14] Similarly, incidents of anti-Semitic harassment, vandalism, and assault climbed 86 percent in the first quarter of 2017 compared with the same period the year before.[15] Additionally, 2017 was the deadliest year on record for the LGBTQ community in the United States, with an 86 percent increase in single incident reports of homicides of LGBTQ people compared with 2016.[16] Transgender people are at particular risk, with twenty-nine killed in 2017, the most ever recorded, and twenty-six by November 2018—before the year was even over. A firearm was involved in more than half of these incidents.[17]

This is why Moms Demand Action volunteers fight against dangerous legislation that would leave marginalized communities at even greater risk of experiencing gun violence, such as the Concealed Carry Reciprocity Act of 2017—a reckless policy that would gut state gun laws regarding who can carry hidden, loaded guns in public. This law would make it easier for people with a history of hateful and violent behavior to carry hidden, loaded guns in public across the country and would leave people of color, LGBTQ people, and other minority groups even more vulnerable to targeted attacks with guns. To make matters worse, it also opens the door to more school shootings, as it would grant permitted gun owners the right to carry concealed weapons in school zones in any state, regardless of state laws.

This is not an America I want to live in, or that I want my kids to grow up in. This law isn't even about maintaining the status quo; it's about reverting the country back to a twisted, retro version of the Wild West. This is where our cause gets even bigger, and more important. It's not just about keeping our kids safe; it's about creating an America that lives up to its ideals of being a land of equal opportunity—and that includes equal protection from the risk of being murdered.

Since January 2017, when the Concealed Carry Reciprocity Act of 2017 was introduced, Moms Demand Action volunteers placed calls and collected and delivered thousands of postcards to their legislators opposing the bill. They attended town hall meetings and met with their elected officials. Despite these efforts, the bill did pass the House. But all that grassroots activism still paid off: 18 House members who had voted for the 2011 version of the bill voted against it in 2017; and in 2017, the bill passed by only a 33-vote margin, down from a 118-vote margin in 2011. This erosion of support essentially labeled the bill dead on arrival; a Senate version of the bill never made it to the floor for a vote. It's just another example of the power of losing forward.

Expanding Our Tent Through Collaboration and Support

Of course, we have to do more than just focus on legislation to heal the uniquely American crisis that causes black and brown people to be more vulnerable to gun violence than white people. We have to fight the systemic racism that contributes to it, too. After all, gun violence doesn't exist in a vacuum; it's related to all kinds of issues—poverty (gun homicides are most prevalent in racially segregated neighborhoods with high rates of poverty),[18] mass incarceration, a biased justice system. We also have to address gun violence in all its forms, including police shootings. And perhaps most doable and most important, we have to make sure that we're not an organization of white ladies pushing for what we think black and brown communities need.

We realized early on that the women who made up Moms Demand Action looked a lot like me—white and middle-aged. The fact that our membership was so unrepresentative of the people who are most affected by gun violence was morally wrong and strategically wrong, because if we truly wanted to move the needle on gun violence, we needed the perspective of people who face it at much higher rates. It was clear that we needed to work hard to have a more diverse staff and membership.

I owe a debt of gratitude to Lucy McBath for helping Moms Demand Action become more diverse, equitable, and inclusive. As a black woman, Lucy couldn't help noticing that when she was performing her spokesperson duties for us, she was speaking mostly to white audiences. So she wrote me a very thoughtful and frank letter in which she said, "Even though I love and respect Moms Demand Action and am gratified to be speaking to so many people, we aren't reaching the black community, and we aren't speaking to the faith community, and it is bothering my spirit because to change the culture, we have to be speaking to everyone." She also shared some of her ideas for how we could speak to more diverse audiences, and I am so thankful she did. I know it took a lot of courage. "I prayed when I sent that letter," she recalls. She shouldn't have been worried, though; I was grateful to receive it, and it was the perfect opportunity to ask her to join our staff and help us build the structures that make a bigger tent.

We've made it a priority that the members and leaders of Moms Demand Action are inclusive and representative of all the different communities that are each uniquely affected by gun violence. For example, since 2017, more than 40 percent of the employees we've hired are nonwhite. A guiding principle we've used to help us in our efforts to diversify is to set what's known as SMARTIE goals. You may have heard of SMART goals—Specific, Measurable, Actionable, Results-focused, and Time-bound. The *I* and the *E* in SMARTIE stand for Inclusive and Equitable. A SMART goal would be to say we want to hire twenty leaders in the next ninety days. If we did that, we'd probably end up with twenty leaders who look a lot like the leaders we already have. So we set a SMARTIE goal to hire twenty leaders in ninety days, five of whom speak Spanish.

We also make it a point to hold implicit bias trainings at every level of membership, from our leaders down to our volunteers—otherwise, we'll keep growing a volunteer and leadership base that continues to look a lot like the middle-aged white women who initially got our organization off the ground. It can be difficult to detect bias in your own actions, and to do so takes a willingness to admit that you've gotten something wrong. You may not even realize that you're reaching out only to people who look a lot like you do. Especially when you're a volunteer doing advocacy work in a limited amount of time, it's all too easy to text only the people you have a personal connection with to invite them, and only them, to a meeting. You may not be intentionally excluding people from the group, but the result is still the same.

Making room for more people under our tent requires a specific mindset to be successful. Lucy has been instrumental in helping us learn how to approach working with black communities. "For too long," Lucy says, "white organizations have been coming into communities of color and acting like the savior—bringing their money and resources and dictating, 'This is what we're going to do for you,' without even learning about what is happening and asking what is needed." So we always seek to listen first.

As Beth Sprunger, a volunteer from Indianapolis, puts it, "We are not here to save, we are here to support." That means not charging in with an agenda, but rather asking what we can do to help, and then taking the direction. "Taking a back seat can be hard when you are used to running the show," Beth admits. "But letting the people in the affected communities tell you what they need and then making it happen is the only way to foster a productive relationship."

Allison Mayne Peters, a volunteer from New York City, says it's helpful to have a point person in your group take the lead on fostering connections to other groups. "We have one volunteer who attends every Gays Against Guns meeting and is our main point of contact with the LGBTQ community; another who focuses on reaching out to communities of color." Having a point of contact makes interactions more personal. Of course, you also want that point person to bring other members to any events so that the connection grows. That's why Allison says, "We show up en masse to any responses to shootings that happen in neighborhoods of color and to events that our point person puts on our radar." When gun violence survivors have to attend the trial of their family member's shooter, the local Moms Demand Action chapter arranges a "court support" team that attends the trial every day in a show of solidarity and to remind the judge and jury that this is about more than just one shooting death. Showing up is a strategy that pays off. "We have seen so many people from the communities we support come to our meetings, phone bank, and canvass," Allison says. It's proof that we are definitely stronger together.

One thing we've found is that we continually need to be on the lookout for more populations that we have inadvertently been insensitive to; it takes continual training to raise your awareness and a willingness to observe, listen, and admit when you've gotten something wrong. For instance, Justine Emily, from the Washington state chapter, says: "I never gave the accessibility of our meetings any thought until I met a volunteer with a visible physical disability. Now I make sure all our meeting locations are ADA compliant."

Another often overlooked demographic in any movement is people in their seventies or older. Jenny Stadelmann of the Illinois chapter reports that it took several months for her chapter to accommodate a passionate group of grandparents. Finally, the chapter held a new member meeting at the clubhouse of a retirement community; forty seniors showed up. "It was one of the most engaged, informed, and motivated groups of any people at any meeting I've ever attended," Jenny says. Since then, the local group leaders have started calling these members (since many of them don't use social media or even email) and arranging carpools to the monthly meetings (since many no longer drive). "It has really provided more diversity to our group."

Diversity doesn't happen overnight, though, and it's not set-it-and-forget-it, either. After our massive influx of post-Parkland volunteers—many of whom were women who look like me—we had to start our diversity work all over again. In truth, our work to be more inclusive and diverse will never be done; if we don't keep focusing on it, at some point we'll look up and realize we've gone backward.

Reaching Out Across the Aisle

Our efforts to build a bigger tent apply not just to volunteers, but also to lawmakers. After all, to change laws, you have to change lawmakers. You can do that in two ways: vote them out or change their minds. Frankly, it will take too long to vote out every legislator who is opposed to any kind of commonsense gun laws. And since our goal is to save lives, we don't have time to wait.

This means that we have to consciously engage lawmakers

who have a track record of supporting the NRA's agenda. And to do that we need to keep a door open to people on the other side of the issue with the expectation that they'll walk through it.

This is not just a hope or a wish. I've seen so many people—regular folks, corporate leaders, legislators—staunchly opposed to any kind of gun regulation change their minds and start supporting our cause. If we wrote off everyone who appears to be our enemy, that conversion wouldn't happen nearly as often; we'd just stay entrenched on our respective sides.

The fact is, politicians *do* change their minds. Tim Ryan, once an NRA A-rated Democratic representative from Ohio, started rethinking his position on gun safety after the Sandy Hook shooting. Then, in October 2017, after the shooting at a country music concert in Las Vegas, he donated $20,000—the same amount of the campaign contributions he had received from the NRA—to gun violence prevention groups, including Everytown. That same month, Minnesota state representative Tim Walz followed suit and donated the $18,950 he'd received from the NRA to the Intrepid Fallen Heroes Fund and pledged his support for universal background check legislation and his opposition to laws that would make it legal to cross state lines with a concealed weapon. Even Kirsten Gillibrand, the Democratic US senator from New York who has made a name for herself as a leader of the progressive resistance since Trump's election, once earned an A-rating from the NRA as a US representative from a rural district in upstate New York (she's more recently earned the NRA's F-rating—good for her!).

The winds of politics also change direction. As recently as 2012, the Democratic National Committee advised candidates running in rural districts to show in their campaign ads photos and footage of themselves handling guns.[19] Fast forward to now, and Democrats are making commonsense gun regulation a key part of their platform. Take, for example, Arizona Democrat Ann Kirkpatrick, who received an A-rating from the NRA during her successful 2010 bid for a congressional seat; during her campaign, she talked openly about hunting with her grandfather. In 2018, she talked about how she gave away those hunting rifles, and she came out as a supporter of universal background checks and an assault weapons ban.

Of course, we still work hard to replace those lawmakers who don't support laws designed to reduce gun violence with candidates who do. But it's often a faster path to work with current lawmakers and seek to build relationships with them via in-person conversations and to use phone calls, emails, and social media posts to let them know that we're watching and will hold them accountable for their votes.

Many times you just don't know where lawmakers truly stand until you engage them in conversations. That's what some new Moms Demand Action volunteers from Massachusetts discovered when they scheduled a meeting in 2018 with their local Republican lawmakers to discuss the benefits of a red flag bill that was winding its way through the state legislature. At the meeting, the volunteers pointed out the many ways that red flag laws have been shown to save lives. And to their surprise, the lawmakers said they would support it. It turns out, these legislators had been assuming that their constituents were happy with their position on gun laws, because they hadn't heard from any voters directly. After that meeting, the bill passed and was eventually signed into law (and by a Republican governor, no less!).

Consider that these were new volunteers who had joined Moms Demand Action only a few months earlier after the Parkland shooting. They weren't seasoned negotiators; they didn't know how to play any political games. They simply showed up and asked.

Fred and Maria Wright, whose son Jerry was killed in the 2016 Pulse nightclub shooting in Orlando, Florida, have been incredible assets in demonstrating how to work with politicians of all affiliations. Fred and Maria live outside of Miami; they're both moderate Republicans and lifelong Catholics. Maria grew up with guns in her home, and she and Fred are both proponents of the Second Amendment; but even before their son was murdered, they supported stronger background check laws. After Jerry was killed, passing stronger laws designed to keep guns out of the hands of dangerous people became their number one focus.

The Pulse shooter had been investigated twice by the FBI on suspicion of being linked to terrorist organizations, but he was still able to legally purchase the guns he used to kill forty-nine people—a semiautomatic rifle and a handgun—only a week before he attacked.

"After our son was killed, we said, 'This is our calling,'" Fred recalls. And Maria adds, "We knew we had to do whatever we can to help people see that it can happen in any family, to make it so that other parents never have to go through this." Within a month of the shooting, Fred and Maria were in Washington, DC, meeting with lawmakers and speaking to crowds.

Fast forward to now, and Fred and Maria are the legislative leads in Florida for Moms Demand Action. In their roles, they meet with politicians at all levels, of all parties. In Florida, "the

NRA's petri dish," as Maria refers to it, they have their work cut out for them. Their position as parents whose son was killed by gun violence and the facts that they are Republicans, Christians, and Hispanic make them uniquely suited to be able to speak with a wide variety of people on the issue of gun violence; but they have lessons to share for those who want to know how to have constructive conversations with politicians on the other side of the issue.

The first is to keep the conversation on a human level and not view it as a battle. "When you go in ready to attack, that makes it very hard to hear each other," Maria says. She and Fred show respect from the get-go by being polite to the lawmaker's aides when seeking to arrange a meeting, and dressing up when they do meet. "We always walk in with a photo of Jerry," Maria says. "We put it on the table as soon as we start. We tell them a little bit about him to show that we're not there for a specific political agenda. Then we say we can no longer protect our child, but we don't want other families to go through what we went through. At the end we always, always thank them for giving us the opportunity to honor our child."

"Usually, even though we might not get to agreement, we do get to conversation," Fred says. Yet sometimes they surprise even themselves with the impact they have.

When the US House of Representatives was due to vote on the Concealed Carry Reciprocity Act, Fred and Maria set out to have conversations with their local representatives, Ileana Ros-Lehtinen and Carlos Curbelo. They began by reaching out to the reps' aides ("Be really, really nice to the aides," Maria counsels), who then worked to set up in-person meetings. It was proving more difficult to find a time to meet with Representative Ros-Lehtinen until one of her aides revealed that she'd be attending a party at a friend's house. Maria recalls: "Fred and our daughter literally crashed the party and talked to her about the bill in front of her friends and husband, explaining how the bill would make the weakest gun laws in a given state the law of the land. Everyone at the party was saying, 'Ileana, you cannot vote for that!'" They also brought groups of Moms Demand Action volunteers to a meeting with Representative Curbelo. And both of the legislators voted "no"—the only two Republicans from the South to do so. The bill still passed the House, but the advocacy of Fred and Maria has helped send the message that this isn't the slam dunk the NRA expected it to be.

Fred and Maria have even made inroads with people when it seemed like there would be no hope of finding middle ground—people like Florida senator Marco Rubio, an NRA A-rated Republican. "He has met with us face to face, and he responds thoughtfully to our emails," Fred says. "We've at least gotten him to discuss magazine capacity, red flag laws, and other issues, where before he wouldn't have even thought about it." One important place where Fred and Maria have agreed with Senator Rubio is on the need for more research on gun violence (after the government banned the CDC from studying it by denying their funding).

But Fred and Maria won't stop there. "We'll soon ask for another face-to-face meeting where we can discuss several things with him and find that middle ground where we can all be safer without taking guns away," Fred says. "We're going to keep trying."

Working with Our So-Called Enemies

Because our country is so divided, partisan politics is one of the biggest barriers to building a big tent. But being inclusive means we need to be able to find common ground with people who don't see things the way we do. Our efforts to diversify Moms Demand Action and raise awareness around implicit bias have been met by our volunteers with open arms. What's been a harder sell is actively seeking to work with people who don't fit the progressive liberal ideological mold.

We have supported plenty of candidates—with donations, activism, and Gun Sense Candidate distinctions—whom many of our predominantly Democratic members strongly oppose. In 2016, we endorsed Pat Toomey, a conservative Republican senator from Pennsylvania who frequently votes to curtail many of the things that progressives hold dear—including reproductive rights, environmental protections, and the Affordable Care Act. But when he stuck his neck out to co-sponsor the Manchin-Toomey bill to close the loophole on background checks, we promised he would have our support when he ran again for office. That was not a popular move among some of our Pennsylvania volunteers. But as long as Toomey stands with us on gun violence prevention, we will stand with him.

A particularly controversial incident happened in 2015 when Lucy McBath appeared in a documentary called *The Armor of Light* with the Reverend Rob Schenck, a fixture of the political right and staunch opponent of a woman's right to choose. In the early 1990s, Reverend Schenck was part of Operation Rescue, the anti-abortion group that protested outside abortion clinics and tried to block women's access. He was even arrested once for confronting women with an aborted fetus. He's also an evangelical Christian—many of whom believe that the Bible supports unrestricted gun rights. One look at his résumé would likely convince anyone that he would never be someone whom Moms Demand Action could consider an ally.

Then, in 2013, Schenck witnessed a mass shooting in his own neighborhood in Washington, DC, where twelve people died in an attack at the Navy Yard. Shortly after, he met Lucy (who is also an evangelical Christian), and, as he recalled in an NPR interview, he "saw her heart, her eyes, the pain in her mother's soul."[20] While I'm sure there were other events that led to Reverend Schenck's changing his thinking about guns, meeting Lucy caused him to challenge his views. He struggled to reconcile his, and his church's, pro-life beliefs with simultaneously supporting a proliferation of guns, and unfettered access to them.

The Armor of Light documented Schenck's efforts to get the evangelical community to reconsider its stance on guns alongside Lucy's struggle to process her grief and advocate for commonsense gun laws. Having a visible Moms Demand Action spokeswoman collaborating with someone with such conservative (and anti-choice) political views did not go over well with some of our volunteers. In fact, we lost some of the volunteer leadership of our Nashville group because of it. And although I understand why they were upset, we are an organization dedicated to reducing gun violence. Reverend Schenck, for all his other views that we may not agree with, shares that dedication. He's a conduit to a population that is generally uninterested in our views, and for the access he provides, Moms Demand Action is thankful.

Stephanie Mannon Grabow, a current member of our national training team and former chapter leader in Indiana—a very red state—is a great example of how our cause has reached out to faith communities. Stephanie is a woman of faith who grew up with guns in her home, enjoys sport shooting, and has a history of being a moderate Republican—she was a staff member for a Republican state administrator, is a graduate of a Republican women's leadership program, and ran a congressional campaign for a Republican candidate. In her work to grow Moms Demand Action in Indiana, Stephanie has reached out to religious communities and asked whether she could speak to the congregants. "Meeting in a place of worship helps neutralize the political discussion and has given us a way to engage people at a heart level and a moral level that we might not have been able to do if we'd gone into the public library and talked about facts and figures," Stephanie says. "If they're regularly attending services, they already believe that showing up is important. We engage them and say, 'You can show up over here too.' That's really been our secret sauce" to growing an engaged membership in a red state.

Stephanie has also been a leader to us in how to bridge the red state–blue state political divide. Like me, Stephanie became engaged in reducing gun violence after Sandy Hook. At the time of the shooting, her son was also in first grade. "I remember that day so well," she recalls. "I still can't talk about it without crying." Stephanie saw my Facebook group, but she didn't join right away. She was, after all, working full time, traveling a lot for her work, and raising a young child. It wasn't until the NRA held its 2014 national convention in Indianapolis and an acquaintance invited her to attend the gun lobby's women's leadership luncheon that things changed for her. "This woman assumed because of my background and political work that I was interested in supporting the NRA," Stephanie remembers. "I was mortified that she thought that I would want to have anything to do with the organization that helped create the conditions that led to Sandy Hook. And I knew in that moment that my silence had been interpreted as support. I decided that I was never going to let anyone ever wonder again where I stood on this issue. I joined Moms Demand Action on that day."

In the early days of Moms Demand Action, Stephanie had to help us recognize how our language sometimes made us culturally unwelcoming to conservatives or Republicans; saying things like "We're going to turn Indiana blue!" would turn off a conservative, who would never be seen again. It's far more effective to stick to the issue—to say instead, "We're going to get more gun-sense candidates elected!"

When it comes to building bridges with Republican lawmakers and conservative community leaders, Stephanie says she tries to bring a survivor with her who can share a story that relates to that leader's constituency. "When we met with the assistant chief of police in Indianapolis," Stephanie recalls, "we brought a domestic violence survivor whose ex-husband and son had both killed themselves with guns and a mother whose child had been devastatingly injured by a stray bullet. We'd been trying to get a meeting with the police for months, and I couldn't even get my phone calls returned. When we finally got in the room with him, he ended up meeting with us for forty-five minutes, then later got us in to meet the chief." Now the Indianapolis group is partnering with the Indianapolis Metropolitan Police Department to distribute gun locks to the community. "It's a slow growth process that pays off in big ways."

Jenny Stadelmann, who lives in the northern suburbs of Chicago where the numbers of Republicans and Democrats are about equal, says her local group started to be more careful about the language they used at Moms Demand Action meetings after a diversity, equity, and inclusion training for our volunteer leaders. "I began starting each meeting with ground rules that emphasized that we are nonpartisan, that we will disagree on many political issues within our group and that's okay, because we are brought together for one very important reason: gun violence prevention," Jenny says. After instituting the new ground rules, Jenny reached out to some folks who identified as Republican or Independent and had attended previous meetings but hadn't returned. Many came to another meeting, noticed the change in climate, and reported that they felt more welcome. "It takes some redirecting when people at meetings wade into other political territory," Jenny says. "I encourage them to take those issues to other groups and focus on what we do agree on, which is a lot."

I'm sure you won't be surprised to learn that many Moms Demand Action supporters identify as Democrats, but I think you might be surprised to know how many of them don't—we're also Republicans and Independents and Libertarians, and sometimes a mix of all the parties when you break things down issue by issue. We come together primarily to do gun violence prevention work, but the outcome of doing all our work to be diverse, equitable, and inclusive is to be able to bring people together who are very different and work together on a shared goal around a shared value. It's the opposite of what the NRA is trying to do, which is to pit Americans against one another. We're seeking any amount of common ground we can find in any number of different groups of people to work toward something that benefits everyone. It's not just what we need to reduce gun violence; it's good democracy. It's also exactly what we need to pull ourselves out of this us-versus-them downward spiral.

10

Let This Mother Run This Mother

Once volunteers and gun violence survivors join Moms Demand Action, it doesn't take long for them to realize that many of the lawmakers who represent them aren't exactly rocket scientists. The more time you spend at city council meetings or statehouse hearings, the more you realize that our nation's patchwork quilt of laws has been sewn together mostly by men who are more interested in power and perks than policy.

I know because I've seen it myself, and it's fair to say I wouldn't trust all that many of them to make me a cup of coffee, let alone make the laws responsible for ensuring the safety of my family and community.

Amber Gustafson, a Moms Demand Action volunteer turned candidate—she ran for the Iowa Senate in 2018, a race she unfortunately lost (although she has her eye on another seat that will be opening in 2022 and is currently working on nonprofit boards to find a home for the fundraising and networking skills she honed on the campaign trail)—had a similar realization: "I always assumed that elected officials were super smart people with tons of skills. Once I started getting involved with Moms Demand Action and spending time at the capitol, I realized how many of them simply did not have their crap together and should not be making decisions that affected other people."

There are also plenty of lawmakers who, regardless of how intelligent they are, clearly don't care about their constituents' concerns. Christy Clark, a Moms Demand Action volunteer who ran for and won a seat as state representative in North Carolina in 2018, recalls how her work as a gun violence prevention activist led to her meeting people from both sides of the aisle: "Some were wonderful and great, and some just weren't—they didn't care if they received phone calls from ten thousand of their constituents asking them not to support a gun bill; they voted for it anyway. On top of that, some of them were mean and threatening to our volunteers—certainly not the kind of people you would want running your state."

Fred and Maria Wright, whom we met in the last chapter and whose son Jerry was killed at the Pulse nightclub in Orlando, were shocked to discover just how little most legislators knew about the bills they were voting on. "In our advocacy work, we've learned that when a bill is introduced, most lawmakers get a party-line memo telling them how to vote," Fred says. "We've learned to ask lawmakers if they know about different aspects of the bill, and they don't have a clue because it wasn't in their party summary." Not exactly the kind of thoughtful leadership you'd hope for from an elected official!

The Bad News About Running for Office as a Woman

Part of the reason our legislatures are subpar is because there are far too few women working in them. As of January 2019, 76 percent of members of the US Congress were men, along with 72 percent of state legislators,[1] 82 percent of governors,[2] and nearly 78 percent of mayors (of cities of more than 30,000 residents).[3] These numbers are abysmal.

But we've made progress: Six times as many women are serving in state legislatures today than in 1971. In 2019, the nationwide share of female legislators is around 28 percent, a seemingly small yet still significant jump of three percentage points over 2018.[4] Thirty-seven percent of those congresswomen are women of color. That said, we can't sit back and wait for the passage of time to move the needle for us. We've got to be proactive. We've got to run for elected office.

This all begs the larger question: Why is there so much gender disparity in government? Researchers have found several reasons for the lack of women in elected office, including the fact that they're less likely than men to be *encouraged* to run—even by their parents and teachers. And unlike men, women underestimate their own abilities and qualifications to hold office. Career and family obligations also factor in, which may explain why the average age for a woman to run for office in the United States is about fifty—when her children are likely to be grown or nearly grown.[5]

Catherine Stefani, a California Moms Demand Action volunteer and the District 2 city and county supervisor for San

Francisco, had the typical background to become an elected official: she was an attorney who had worked as a legislative aide for years and had completed a training for aspiring female candidates back in 2009. She also had always known that someday she wanted to run for office, but still she didn't do it until after the mayor of San Francisco died. As a result, her boss, a city and county supervisor, was appointed to become mayor, and, in turn, Catherine was appointed to fill her boss's shoes. In 2018, she ran to keep her seat, and she won. As qualified as Catherine was, it took a lightning strike to get her to run.

Even Lucy McBath, who had been a nationally recognized gun violence prevention advocate for years, says no one suggested she run for office (in 2018, she became the congressional representative for her district in Georgia) except for Moms Demand Action. "There was no one coming to me saying, 'We need your voice,' until Shannon started saying things like 'When you run for office . . .'" Even after years of loving nudges from me, it took the 2016 election and a hip replacement—which forced Lucy to sit still long enough to deeply consider her next move—to persuade her to run. "I prayed a lot during the winter of 2016, and then the pieces started to fall into place," she says. The political action committee EMILY's List and then the Democratic Party of Georgia gave Lucy awards for her advocacy. Then Georgia state representative Renitta Shannon took Lucy out to breakfast and encouraged her to run. It took all these events for her to realize that she had more than enough experience to say "yes."

And really, it's no surprise that it took so many years and so many people to influence Lucy: a 2013 study found that even among today's generation of college students, men are twice as likely as women to have considered running for office someday. That same study found that men are 15 percent more likely to be recruited to actually run.[6]

And once they do decide to run, woman candidates still face more obstacles than men. Research shows that the media treat female candidates differently from male candidates—they're covered less, and when they do make the news, the coverage is often focused on their appearance or their likability.[7] Sound familiar? Unbelievably, a 2015 study found that some voters prefer male candidates even when evidence clearly indicates that the female candidate is more qualified.[8]

Women also have to overcome gender bias among voters or donors who don't take woman candidates as seriously as men. "As a woman, and especially a woman of color, I have to raise twice the amount of money because it's harder to get people to believe in how I can make a difference," Lucy McBath says. In addition, raising money is harder for women because, as newer candidates, they typically don't have the relationships that provide access to money that white male candidates have.

And on top of all that, women are held to a different standard regarding their looks. "I have had people ask me, 'What are you going to do about your hair?' Or your nails, or your voice—I have a quiet voice. These are things they would never tell a man," says Christy Clark.

Looking at all the obstacles we have to overcome, I can understand why women resist the idea of running. But the environment isn't the only sector experiencing climate change—so is the political realm, and the tide is turning toward women. And honestly, given where this nation is right now regarding its treatment of women, especially women of color and marginalized communities, running isn't just a nice-to-do—it's a moral imperative.

Having more mothers in office is also how we start to do better on the issues that affect children and families the most. Rhode Island governor Gina Raimondo is a great example of this. "My priorities are absolutely influenced by my role as a mother," she admits. "I've made record investments in education and I've really leaned in to gun violence and health care. I've seen my kids flourish because of good pre-K and full-day kindergarten and access to good health care, and I want the same for all kids." By running for office, you don't just give women a seat at the table; you also bring your and everyone else's kids along with you.

The Good News About Being a Woman and Running for Office

There are women who are starting to realize that you don't need a perfect résumé to hold office. Jennifer Lugar, a Moms Demand Action volunteer from Jenkintown, Pennsylvania, whose husband shot and killed himself in 2009, is one of them. Jennifer had been an activist for years, but it wasn't until she stepped up her involvement during the 2016 presidential election that the idea of running for office became real in her mind. "Very quickly the mystery of politics started to erode," she recalls. "When you start showing up, you start knowing the people and spending more time in the campaign office. You see how accessible it is." Then, when a reality TV star who had never held an elected office before won the presidency, "this huge barrier in my mind was removed—I couldn't use the excuse that I wasn't qualified enough anymore because I was at least as qualified as the president."

In December 2016, Jennifer received an email saying that the borough council of Jenkintown had two openings. "I didn't even think about it for more than two minutes; I attached my résumé without even reading it over first and threw my hat in the ring." Nine people were interviewed for the spot, and in January 2017, Jennifer was appointed. In November 2017, she ran to keep her seat, and she won.

Despite the unique hurdles woman candidates face, research shows women are as likely as men to win when they run for office (perhaps because they run only when they are ridiculously qualified) and that voters put more weight in a candidate's party, not their gender, when making their ultimate decisions.

And data show that once they get into office, women may make more effective elected officials for their constituents than men: a study published in 2018 by researchers at Georgia State University found that female legislators secure more federal funding for their districts and introduce more bills and resolutions than male legislators do. Women also do the work it takes to stay in office once elected: these same researchers found that congresswomen send 17 percent more snail mail communications to their constituents than congressmen do and are more likely than their male counterparts to take on committee assignments that reflect the interests and demands of their districts. Furthermore, women are more likely to vote in ways that reflect their constituents' needs.[9]

As Michele L. Swers, a professor of government at Georgetown University who studies gender and policymaking, told the *New York Times*, "All members of Congress have to follow their constituency, but because of their personal experiences either as women in the work force or as mothers, [congresswomen] might be inclined to legislate on some of these issues."[10]

Our naturally high empathy levels are a huge piece of what makes us more effective elected officials. "Democracy depends on being able to connect with the people you represent," says Jennifer Lugar. "That's a characteristic that is too often missing in male lawmakers, and something that more women have. Because we take care of so many people and different facets of life, we have a more holistic perspective."

Woman lawmakers have also historically been more aggressive in addressing gun violence through policy and legislation than their male peers. In 1994, California senator Dianne Feinstein was the architect and chief sponsor of the Assault Weapons Ban. When it came to a vote, 83 percent of the women in the House, and all but one woman in the Senate, voted in favor of the ban; only 50 percent of the men in the House and 59 percent of the men in the Senate voted for it.[11]

More recently, Minnesota senator Amy Klobuchar and Michigan representative Debbie Dingell worked together to craft legislation that would keep guns away from domestic abusers by broadening the definition of prohibited purchasers to include dating partners and stalkers convicted of abuse. For both women, the issue is personal. As a county prosecutor, Klobuchar helped establish some of the first domestic violence service centers in the nation. Dingell grew up in a household where domestic gun violence was a constant threat. As I write this, the bill has yet to make it out of committee, but these women have continued to put it on the docket each year since

2015, the year they first introduced it. I have no doubt that their resilience and dedication to protect other women and children from violent domestic partners will ultimately result in a win.

The NRA, which has fought against *all* major gun legislation for decades, also opposed these bills by woman lawmakers—and, not surprisingly, it has labeled a disproportionate percentage of the woman lawmakers in Congress as being "anti-gun."

But the NRA can't intimidate and control women as easily as they can men, because we're driven by something higher than simply staying in office for the sake of staying in office. Christy Clark says, "Moms aren't thinking only about ourselves and our goals; we think of our children and grandchildren as we make decisions, and that's what separates a male politician from a female politician. And we're not trying to make life better for just our own kids, but for all the kids in every neighborhood, school, and home." In other words, we're not motivated by a love of power; we're driven by a love for our kids and a desire to create a country that's a safe place for them to grow up. There is no greater force in the world.

It's a force that fueled Nikki Fried, a former commercial-litigation lawyer and public defender, to run for—and win!—the office of the Florida commissioner of Agriculture and Consumer Services, which presides over the gun-permitting process. Nikki says she was inspired to run—and to keep going—in large part by Moms Demand Action volunteers. "A campaign trail can sometimes be a lonely place," Nikki admits. "Any time I needed support and encouragement, the women of Moms Demand Action were there. Their presence was a constant reminder of why I was running—to help these hard-working mothers who are fighting for children all across the country who need people like me to step up and help protect them."

Nikki's predecessor, Adam Putnam, had been hugely influenced by NRA lobbyist Marion Hammer, who would practically dictate changes she wanted to see in the permitting process.[12] Worse yet, Putnam was found to have neglected for thirteen months the background check process for people requesting concealed carry permits, enabling more than eight hundred people who shouldn't have been eligible to legally carry a concealed weapon. During her campaign, Nikki promised very clearly that if she was elected, the NRA would have no influence in her department. "My job is to have background checks done properly," she says. "We're not going to be cutting corners for time efficiency, we're going to do it right." Her win is a huge step forward for gun-sense policies, and a perfect example of why we need women to run for a variety of political offices so that we have access to many levers of power when it comes to making America safer from gun violence.

These woman lawmakers who are working so hard to strengthen our nation's gun laws are simply reflecting the will of American women at large. Polls continually show that women support gun safety more than men. Recently, a Quinnipiac University poll found a "wide gender gap" in views on gun laws: 69 percent of women polled supported stricter gun laws, and 26 percent didn't; while the split was down the middle for men—47 percent supported stricter gun laws, and 47 percent opposed them.[13]

As I write this, there will be 127 women in the 116th Congress[14]—35 new women were elected to the House of Representatives in 2018, a record that makes us much better positioned to make women's desire for stricter gun laws a reality. I can't wait to see the gains we'll make together.

If You Want to Win, You've Got to Train

About two years after I founded Moms Demand Action, it became clear to me that our volunteers were looking to take the skills they'd learned as activists and apply them to running for office. I started seeing volunteers announce on social media that they were exploring a run for school board or city council or had already filed to run for office. So I went to our leadership team and suggested that we develop a more formal training program.

Moms Demand Action was already providing our volunteers with training for a variety of advocacy-related work, including communications coaching, guidance on growing membership, working with data, etc. So why not add a module that would encourage and prepare our volunteers to run for office? If nothing else, it would plant the seed in the minds of volunteers and gun violence survivors that becoming an elected official was the next logical—and completely attainable—step on their paths as activists.

Soon after, we created a comprehensive guide to help volunteers navigate tough questions when deciding whether to run for office. Developed in conjunction with experts in political campaigning, the guide helps volunteers zero in on which race they should choose to enter given their passions and geography. It also provides important information about the key components of a successful campaign, from creating a month-by-month calendar to finding funders.

Furthermore, we've created training modules to help potential candidates drill down on campaign specifics, including the creation of an "elevator pitch": a personal story about their campaign that can be told in the length of an elevator ride. Another module explains how to create a budget that will carry candidates through their campaigns, and how to raise money. And another explains how to energize and mobilize voters, including specific directions on how to organize a canvass.

In addition to this training, we urge our volunteers to seek out local organizations that focus specifically on training women for office. For several years I've been on the national board of the nonprofit organization Emerge America, which seeks to increase the number of Democratic woman leaders, and I graduated from its Colorado program in 2016. (In case you're wondering, while I have certainly learned that life can take you in a direction you didn't expect, I don't have any plans to run for office. I'm confident I can do more good where I am. Here, I get to help create change at every level—local, state, federal, and within women's own minds and lives.)

Emerge America focuses specifically on training progressive women to run for office and has an excellent track record of helping to elect women to municipal and state elected positions. Programs like Emerge augment the training Moms Demand Action offers by broadening the issues they focus on beyond gun safety. They also help women create relationships in communities and with other women who can help support their candidacies. There are training programs for woman candidates of all types: the National Federation of Republican

Women is for, well, Republican women; the Victory Institute focuses on LGBTQ candidates; and the Women's Campaign Fund is a bipartisan organization that seeks to train female-identifying candidates of all ideologies and ethnicities.

Women supporting women is key to any successful race, and once our volunteers do decide to run—and go through our training and file for a specific race—they then have a built-in support system of fellow Moms Demand Action volunteers willing to canvass, make calls, and hold fundraisers for them in their communities.

Amber Gustafson says, "My local Moms Demand Action volunteers were just amazing!" Granted, moms in general, and Moms Demand Action volunteers in particular, are busy, so they often have to find ways to involve their kids in their volunteer work. "They marched in parades with me," Amber says. "It was such an easy and kid-friendly thing to do, and it helped fill a need that not a lot of other volunteers are interested in showing up to do."

Moms Demand Action candidates are great at finding ways for moms to volunteer for their campaigns in a way that makes sense. Catherine Stefani, who, as our former San Francisco chapter lead for four years, has multiple years of experience organizing volunteers, points out that the people who support your candidacy may have reservations about being qualified to help or not having enough time to be truly helpful. The way to cut through that, Catherine says, is to let everyone participate to the level that makes sense for them. "I told my team, 'Give what you can and don't judge it, and I won't judge you.' By giving volunteers the opportunity to work with their schedule and their strengths, you can build up an incredible team" to help you knock on doors, make phone calls, and write post-cards. She used this approach to build the Moms Demand Action chapter and to get herself elected.

In addition to doing the legwork, volunteers support the candidates who rise through our ranks financially, giving from their own wallets but also leveraging their personal networks to spread awareness and raise funds. And that's such an important leg up for female candidates with a history of advocacy; it gives them a grassroots strength that more traditional candidates—particularly incumbents, who are much more likely to be already beholden to corporate or lobbying interests—can only dream of.

Motherhood Is a Political Asset

We've talked a lot about the political constraints for women running for office, as well as the sacrifices and training required to become a candidate, but moms have a real advantage in their races these days. Motherhood has become a potent weapon in modern-day politics, enabling women to convey the authenticity and outsider status many voters are looking for in a post-2016 political climate.

Motherhood has actually long been a part of politics in the United States. Lydia Sayer Hasbrouck, a mother of two, is thought to have marked the first-ever electoral victory for an American woman when she was elected to the board of education in Middleton, New York, in 1880. Susanna Salter, a mom of nine, became the country's first woman mayor in 1887 in Argonia, Kansas.

Flash forward to 1992, when a wave of women were elected to Congress in the wake of the Anita Hill hearings the year before. Patty Murray, from Washington state, campaigned for a US Senate seat as a "mom in tennis shoes" and won the seat when both of her children were younger than sixteen; she still holds the seat in 2018, twenty-six years later. And then in 2008 there was Sarah Palin, the Republican vice-presidential candidate who notoriously cited "lipstick" as the qualifying difference between a hockey mom and a pit bull. A decade later, after Senator Tammy Duckworth became the first senator to give birth while in office, the Senate unanimously voted to allow children younger than one year onto the floor during votes (older children are still prohibited) and to allow female senators to breastfeed during votes. Not surprisingly, Duckworth led the charge on that rule change, proving that the more women who hold office, the more they pave the way for others to join them.

Moms in office know that being a mother improves their work—from more acutely understanding the needs of children to providing a real-life perspective on policy decisions related to women and families. They also take moral action in a way that other politicians typically haven't. Which is why more and more candidates are emphasizing their "momness."

Asked how she could be a mother and a member of Congress at the same time, Colorado congresswoman Patricia Schroeder said, "I have a brain and a uterus and I use both." Amber has this to say about what she's learned about the political power of moms on her own campaign trail: "I know there are people who feel that being a mom is a strike against you, but boy I sure haven't found that. If anything, it's a boon."

In the past decade, moms running for office have become

more daring. Their campaign ads highlight the need for protecting children and for high-quality daycare; they even proudly breastfeed on camera. Moms are taking their babies and young children to campaign events and drawing on stories about them to make emotional calls to action.

Sometimes, this daring shows up as admitting to ourselves and to the world that we have a bigger mission and we want to skip over the "starter office" and run for a more powerful position. That's what happened with Lucy McBath, who was initially inspired to run for the Georgia legislature. But then the Parkland shooting happened. "Those children were the same age as Jordan when he was gunned down," she says. "I kept thinking, 'Who's standing up to help? Who's going to support the kids?'" That moment, more than anything, helped Lucy see "that everything I've learned in my work with Moms Demand Action and as a gun violence prevention advocate has prepared me. Now I understand I haven't just been doing grassroots activism; I've been preparing to run and training to lead." That's when she decided to run for the US Congress instead. Because as important as it is that we fill offices at every level of government with women who can bring empathy, common sense, and hard work to the job, it's essential to have them at the top, too.

Even If You Don't Want to Run, You Can Still Be an Electoral Force

Of course, there are many ways to change the face of electoral politics that don't require you to run for office. For example, many of our volunteers find that the skills they've developed and the networks they've cultivated in their advocacy work empower them to be effective supporters of other candidates.

Norri Leder, who spearheaded the Texas signage campaign I talked about in Chapter 4, put her experience with Moms Demand Action to great use in 2018 when she supported Lizzie Fletcher, a gun-sense candidate who ran against Congressman John Culberson, an NRA A-rated incumbent. "I thought Lizzie was such a strong candidate," Norri says, "and voting John Culberson out had been a dream of mine since Sandy Hook—it was a perfect storm for me." So she reached out to Lizzie's campaign and was quickly off and running. Norri put together meet-and-greets for Lizzie at people's homes and did a ton of social media posting on behalf of the campaign. She also enrolled her fellow Moms Demand Action volunteers to make phone calls and write postcards. And Norri played a key role in educating Lizzie about the intricacies of the gun violence issue and helped her develop her policies on gun reform.

In the end, Lizzie won by a margin of more than five points—a definitive victory in a state as historically NRA-friendly as Texas. "It's one of the most gratifying things I've done outside of raising my kids," Norri says. "It's so emotionally rewarding to feel like I'm having a direct impact on this issue." Now Norri has another strategy in her arsenal to effect change—one she plans to use again. "I definitely feel like I am going to return to campaign work. I know that if I pick my candidate carefully, I can add my voice and my elbow grease and help push that candidate over the top."

Another volunteer—and Texan—who got to feel the rush of how much one person's voice can influence a campaign is Diana Earl, a gun violence survivor whose twenty-two-year-old son, Dedrick, was shot and killed in late 2016 (and whose story I shared in Chapter 1). By early 2017, Diana had gotten involved with Moms Demand Action; she dove in by testifying at the statehouse in Austin against a bill that would legalize unlicensed carry. "It was empowering to testify because I knew if I wanted to create change, I had to share my story with people who had the power to create change," Diana says. That empowerment only grew as election season got into full swing over the coming months, inspiring Diana to start following Beto O'Rourke's campaign to replace Texas senator Ted Cruz. And she did more than just read up on him; Diana threw on her red Moms Demand Action T-shirt and went to so many of O'Rourke's town hall meetings (up to five in one day!) that he started recognizing her. "Every time I heard his message on gun reform, criminal justice reform, and immigration reform," Diana says, "I got energized and I knew his was the campaign I wanted to work on."

Diana's experience with Moms Demand Action and with the O'Rourke campaign helped her see how valuable her voice as a gun violence survivor is—a tool she plans to continue to use to help get gun-sense candidates elected in the future. "Campaign work is going to be a focus for me," she says. "I see that as something I'll do for the rest of my life." I'm hoping that one day the campaign Diana chooses to work on will be her own, but until then, I know good candidates will have a powerful ally in her. Every candidate needs a team of smart and passionate people helping them win; and with margins between winning and losing candidates as razor thin as they were in many 2018 races, you can see a clear connection between the work you put in and the final result. Also, getting involved in a campaign is a great way to build your political experience as you ponder a possible future campaign.

Tools for Women Contemplating a Run for Office

For anyone reading this chapter and starting to wonder whether holding a political office might be in your future, here are some of the most important strategies for persuading yourself—and others—that you are the perfect person for the job:

- **Own your experience.** It doesn't matter if your career doesn't include a high-powered position in corporate America. Raising kids is, in itself, a full-time job with a lot of responsibility and authority. If you've served in any positions where you've had to organize people and keep the trains running, you probably already have at least as much experience as your state representative—perhaps more. Some people say running the parent-teacher organization at their kids' school is the hardest job anyone could have—so own it! "Being on the PTO taught me public speaking, how to organize volunteers, use social media, and raise money," Amber says—all crucial components of running a campaign and holding an office. Brag about all the ways in which you've kept your family running, including experience in your kids' schools and your community.
- **Talk about your unique skill set.** Being a mom includes a built-in leadership boot camp. Motherhood turns you into

an expert dealmaker, preparing you to simultaneously stand on principle and work with others on compromise. You know how to balance interests that can compete for attention and resources. And you master one of the secret skills of motherhood—the ability to multitask like a mofo. You're already caring for everyone in your immediate family, and perhaps an older relative, and likely volunteering in some capacity, whether at church or at school; it all counts. "Getting stuff done is getting stuff done, whether it's in your house or in the statehouse," says Christy Clark. Which is why she believes, "If you want something done, ask a busy woman to do it."

- **Acknowledge how adept you already are at managing a budget.** Many women are intimidated by the thought of raising money and managing the financial end of a campaign; the money side of politics can feel like a bridge too far. But you're likely discounting how much experience you have managing money—even large sums of money. According to a survey by Allianz Life Insurance Company of North America, more than half of women say they're the primary decision-makers on financial matters in their households.[15] That's a tough job considering that according to the US Department of Agriculture, it costs $233,610 on average to raise a child to age eighteen.[16] As a result, moms know how to track their spending, rein in expenses, and negotiate a good deal.
- **Break the mold.** Woman candidates are pushing back on how they're supposed to dress and where they're supposed to live—all the trappings that have long been associated with politicians that are quickly dissipating (and none too soon). "On the campaign trail, I don't wear a suit or a jacket

because they're not how I typically dress," says Christy Clark. Christy also lives in a middle-class house, while her opponent lives in a multi-million-dollar lakefront home. "People say to me, 'You're a regular person.' I say, 'That's the point.'" Our ideas of what candidates "should" look like will never change if we don't put ourselves out there for the country to see.

- **Leverage your network.** It's a simple truth that to run for office, you have to know a lot of people. What's great about being a mom is that your kids force you to know a lot of people, and all different kinds of people—fellow parents, pediatricians, coaches, teachers, and caregivers. "The relationships I've made just from being a mom have given me an awesome base of support," Amber Gustafson says. These people don't just know who you are, "they know your values and how you work and treat people." When they hear you're running, they're thrilled to know that someone they trust is stepping up, and they become wholehearted supporters.
- **Just get started.** I hear a lot of women talk about having impostor syndrome—that feeling of comparing yourself to others and pretending you know more than you do—but no politician knows what they're doing at the beginning. "It was shocking to me to see how quickly you can get up to speed and become an integral part of your state's party politics or city government," says Jennifer Lugar. You really can learn as you go; you just have to take the first step, and then the step after that, and the step after that.
- **Rethink what it means to be qualified.** "The people who are in positions of power benefit from the rest of America

thinking that running for and holding office is only available to a select few because it allows them to hold on to the reins of power," Amber says. "You don't have to be highly educated, have a law degree or a million dollars to run. You need good communication skills, a thick skin, and no fear of hard work." Those are basically all the same qualities you need to be a good mom. *Which you already are!*

- **Count on support from your soul sisters.** If you decide to run (and are a gun-sense candidate), you can absolutely rely on Moms Demand Action to support you. We've done it for numerous politicians all over the country, including Oregon governor Kate Brown, who in 2015 signed a law to close the background check loophole on gun sales made via private and unlicensed gun dealers. "You don't expect moms to show up," Governor Brown says, "whether it's to lobby at the capitol or go knock on doors—it's easy to think that moms are too busy. But Moms Demand Action volunteers show up in massive numbers, again and again, and they are fierce." During the run-up to that Oregon law being passed, the NRA and her opponents used every tool in their arsenal to try to block it, including burning dummies of Brown in effigy ("If these extreme groups continue to burn mannequins of me in effigy every time I stand up for the safety of Oregonians, then they're going to run out of mannequins," she said at the time). But knowing that she had the support of so many Moms Demand Action volunteers has helped Brown stay strong. "Having the Moms with me, side by side and shoulder to shoulder every step of the way, provides a level of confidence, faith, and trust that it's going to be okay."

Are You Ready to Jump In?

I was moved when I read a quotation from Jodi Berg, the CEO of Vitamix, who asked the audience at a women's leadership forum: "How many of you were ready to be a mom before you became a mom? Anyone? And yet you did it—the hardest job in the world. . . . So why worry about a job that's not nearly as important as being a mom?"

If you've ever had the idea to run for office—no matter how fleeting, or how serious—what will it take to get you to throw your hat in the ring? For many Moms Demand Action volunteers-turned-candidates, it was a specific event that spoke to them as a mother and inspired them to run. Maybe it was when their kids started having active shooter drills at school, or when a bad gun bill passed in their state. Whatever development makes you realize that the only way to improve our lackluster legislative system is for good people to run for political office—and that includes you—don't push it away. That feeling serves a purpose. It's calling you forward. You can trust it. We women will have your back.

11

Keep Going

On election night 2016, I wore a white dress to the Javits Center in New York City in anticipation of celebrating our nation's first woman—and first mother—to be elected president. I was so excited to have been invited; I couldn't believe I was there as a guest of Hillary Clinton herself! As my plus one, I brought Michele Mueller, an Ohio Moms Demand Action volunteer who had traveled her battleground state tirelessly to campaign for Hillary. (Michele has a knack of being where the action is—she was also present at that new chapter meeting in the Kentucky library with the armed men in attendance, as we saw in Chapter 3.)

We were on a high as we saw one famous person after the next. *Glee* star Chris Colfer stood behind us in line to get into the event, and former secretary of Health and Human Services (and now US representative from Florida) Donna Shalala sat down next to us at our table. The actor Melanie Griffith talked to us about the importance of getting moms to organize as we got snacks from the buffet table. The feeling in the room was electric, even though right from the earliest results the vote totals weren't where we expected them to be. Everyone kept saying, "Don't worry, she'll pull it off." But as the hours went on, the energy downgraded from elation, to concern, to shock. Instead of the glass ceiling shattering that night, it felt like the sky fell.

At about 9:30 p.m., I realized that Hillary was going to lose, as did a lot of other people around me. Many started crying. I started shaking the same way I had in my early twenties when my parents told me they were getting divorced; it was as if I were standing outside in freezing weather with no coat. I knew I wasn't going to be able to keep my composure and that I needed to get out of there. Fast.

As I walked back to my hotel room, shaking so much I could barely hold on to my phone, I called my husband. "What will we do?" I asked him, not sure whether I meant our family, all women, or the nation collectively. I don't remember what he said—I'm not sure I had the capacity to listen to reassurances yet. I just knew I needed to get back to the hotel and try to figure out how I was going to respond to the questions I'd get from the media inquiries I knew were coming. And I had no idea how I was going to do that because I was officially losing it.

When I woke up the next morning after a fitful three hours of sleep and the results were the same—somehow, Hillary Clinton still had lost the election—I began crying and didn't stop. I cried as I checked out, I cried on the ride to LaGuardia Airport, I cried the whole plane ride to Denver, and I cried during the car ride to my home.

In addition to being disappointed for my gender, I was crushed because Hillary had been an outspoken advocate for gun violence prevention. During her campaign, she'd called out the NRA as one of her biggest enemies and had made addressing gun violence prevention a pillar of her platform. She'd traveled with the Mothers of the Movement, which included Moms Demand Action spokeswoman Lucy McBath. I was sure we would finally get the chance to play offense and spend the next four years passing laws that would save lives. And now that promising opportunity evaporated like a mirage. After all, Donald Trump had received $30 million in campaign contributions from the NRA. In fact, he had been the earliest Republican candidate to ever be endorsed by the gun lobby. Not only would we not be winning, but we were going to be playing yet more defense, and to a greater extent than ever before.

The truth is, we hadn't prepared for Trump winning. We hadn't even considered it. The polls and momentum pointed to an easy win for our candidate. Even though the gun violence prevention movement didn't hinge on whether Hillary Clinton was elected or not—ever since Manchin-Toomey we've been focused on effective legislative change at the state level, and we still would be—for a week or two, it felt like maybe it did.

Now it appeared that the NRA had an open pipeline directly into the White House and would be looking for a return on its investment. That fear was amplified a year later, in December 2017, when NRA President Wayne LaPierre showed up at the White House Christmas party, which just so happened to fall on the five-year commemoration of the Sandy Hook shooting. I had attended that same holiday party just a year earlier—along with many gun violence survivors—when it had been hosted by President Obama. The juxtaposition was sobering.

After a couple days of honoring my postelection sense of loss, my husband said to me, "You need to pull it together, be a leader, and make sure everyone feels hopeful. Because that's your job." Begrudgingly, I knew he was right. I needed to dig deep and rely on the resilience muscles I'd been building since the beginning of Moms Demand Action. After all, there's a reason #KeepGoing is one of our most used hashtags in our volunteers' social media posts. It's part of the ethos of our organization, and what we remind each other to do all the time. So I wrote this Facebook post:

> Someone wise (OK, my husband) pointed out to me that wallowing in despondency over the election results is a luxury too many Americans don't have. Including those who have been shot and killed. So if I want to be sad for someone today, it's going to be teens with black and brown skin who will be targeted by armed vigilantes. I'll worry about abused women whose partners have guns and are enabled to kill by weak gun laws. I'll be concerned for children who are at risk of having guns forced into grade schools and onto college campuses. No more posts about existential despair. We all must decide what we're going to do and do it.

Bob Weiss's daughter, Veronika, was shot and killed at UCSB. He sent me this note: "Since Veronika was murdered, I've changed. I'm still grieving. Probably always will but shit doesn't faze me or scare me anymore. I've already survived a fate worse than death. This horrible election result is a temporary detour. We are winning and will continue to. Hillary is our hero. Our job would be a lot easier with her in the White House. Maybe. She would have faced the same roadblocks as Obama did. We can focus on the states during this brief Trump intermission. Hang in there. I'm proud to be with you."

That's the truth: our movement has never been about one election—it's about saving lives. And you can bet that with their champion newly elected to the White House, the NRA will strike while the iron is hot in Washington and in our statehouses. That's why I'll keep waking up every day as a Moms Demand Action volunteer to organize and speak and travel. And that is an HONOR for which I will never be able to repay the universe.

So here I am, picking my armor back up and putting it on. It's super fucking heavy right now. But it still fits like a glove.

Seeking to create systemic change in any area is a huge undertaking that will inevitably have setbacks—some that will fuel your determination to continue, and some that will hit you hard and make you question the wisdom of even trying. Just as motherhood is a job that is never truly over, neither is advocacy. A popular and inspiring quotation from the Jewish tradition says, "Do justly, now. Love mercy, now. Walk humbly, now. You are not obligated to complete the work, but neither are you free to abandon it." Activism and motherhood are both a marathon, not a sprint. As such, they both require resilience, strength, and stamina. Also just like parenting, advocacy will call on you to grow as a person; it requires you to learn how to take care of yourself physically, mentally, emotionally, and spiritually so that you can stay in the game and continually be your best.

Granted, I'm only six years into this ride, and I have no idea when or how it ends. But there are already several skills and tools that I and the volunteers of Moms Demand Action have honed. Some of them are practical, some are more psychological. They help us keep going, and they can help you keep going, too.

Be Devoted to Your Self-Care

When you're doing the work to rectify a systemic problem, fitting it into the cracks of your life and not getting paid for it, it's easy to burn out—especially if you let activism take up so much time that you stop doing the things you need to do to take care of yourself. Self-care is a *huge* component of long-term success. You won't be able to keep going without it.

A lot of people hear the term *self-care* and think it's a luxury, like visiting a spa or getting a massage. But there are all manner of ways to care for yourself; they all matter, and many of them are free. Saying "no" to a request to take something on can be self-care. So can setting your phone to airplane mode for an afternoon, a day, or even a whole weekend. Honestly, the biggest obstacle to feeling good is often just giving yourself permission to stop "getting things done" and instead doing the things that aren't on your to-do list.

Self-care is so important that at Moms Demand Action we've baked it into our organizational structure: we train volunteers on self-care and even host yoga classes, nature outings, and meditation seminars. One of the biggest things we emphasize in our self-care training is that everyone needs to know how and when to say "no" and, just as important, to not feel bad about saying it. To that end, we have a process that allows people to pass the baton when they need to. After all, we know that the majority of our volunteers have families, and there will always be times when family comes first, no matter what you're working on with Moms Demand Action or how pressing it may feel.

For self-care, our members report doing things like reading, bird watching, taking walks in the woods, drinking (non-boxed) wine, eating chocolate, watching *The Golden Girls*, lifting weights, doing yoga, and running. Personally, I try to meditate and exercise every day (I particularly enjoy hiking).

Making space for self-care in your life can help you have the physical stamina you need to keep showing up—after all, it's hard to get to the statehouse if you've thrown your back out after too many hours of sitting at your computer, or skipped so much sleep that you got run-down and caught a virus. Self-care also goes much deeper than that; it provides an outlet for you to clear your head, reduce your stress, and give yourself space to process your emotions. And let's face it, any advocacy work can be an emotional roller coaster—particularly in an area where loss of life is such a big component. Think of it this way: If your kids were dealing with something really hard, would you push them to do more than they were capable of, or would you try to comfort them, support them, and soothe them however you could? Well, sometimes you need to mother yourself as well as you mother others.

Put Your Powerful Emotions to Use

After the 2016 election, the emotion I felt most strongly was grief. The fact that we were so close to having a gun-sense advocate in the White House only to be denied felt like a huge loss. In that instance, my grief felt temporarily debilitating. But it was also grief over the thought of those twenty children and six educators murdered inside Sandy Hook Elementary School that made me decide to start the original Moms Demand Action Facebook page in 2012. While emotions don't generally bend to our will, and they can require that we devote time and energy to our own healing above anything else, they can also be a powerful instigator of change.

Tragically, we encounter grief regularly in the gun violence prevention movement. Every tragedy shakes us as a nation and rips loved ones out of survivors' lives. As a society, we tend to try to avoid grief as much as we can, but gun violence forces us to reckon with it again and again, whether we experience a loss directly or we're mourning for others and for the overall state of things for our kids and our country. It's such a shape-shifting emotion that everyone experiences it differently—sometimes it morphs into rage; for others it looks a lot like depression; sometimes it takes multiple forms within the space of a day.

Like any emotion, grief seeks an outlet for release, and advocacy can be an empowering and cathartic way to facilitate that. Moms Demand Action is often approached by survivors within a couple of days of losing a loved one. Of course we want to support them and help them give voice to their feelings and use their voice to effect change; we also want to refute the idea that immediately after a tragedy is "too soon" to talk about changing gun laws. But we also have to be sensitive to the waves of emotions any survivor inevitably is subject to. We weren't always great at this—we've had to learn a lot as we've gone along, with a ton of feedback from survivors and from Debbie Weir, our managing director, who had so much experience working with grieving people during her tenure as CEO of Mothers Against Drunk Driving. Now we have policies in place that ensure that survivors have psychological support (and we have volunteer psychologists to work with them) before they dive into activism. We have an entire team of people dedicated to supporting and working with survivors, running our survivor support network, and training our volunteers to be supportive of and sensitive to survivors' needs in the short term and the long term.

A frequent companion to grief and another big emotion that can be channeled into action is anger. As I mentioned back in Chapter 1, anger is a loaded subject, particularly for women. If we raise our voices, we get characterized as being hysterical or some kind of fire-spewing monster. And granted, getting angry can make anyone, male or female, do things they later regret. But anger is like inflammation—finite amounts of it in response to a triggering event are purifying, even healing. It's only when anger becomes chronic and your go-to response to any threat—no matter how big—that it becomes a problem.

Rebecca Traister in her 2018 book *Good and Mad: The Revolutionary Power of Women's Anger* writes how women's anger is mostly denigrated and even punished. Think of Serena Williams getting penalized in the 2018 US Open final for defending herself after an unfair call from the umpire, and how she was penalized even more harshly for her response to the initial injustice. But, Traister says, when a woman's anger is related to her status as a mom, it becomes a lot more acceptable. Think of the conservative women who ran for office during the Tea Party uprising in 2010 whom Sarah Palin dubbed "grizzly moms," or even Mary Harris Jones, who advocated successfully for the rights of miners in the late 1800s: she referred to the miners as "my boys" and became known as Mother Jones. There's a long history of a mother's indignation being a force for political change, and there may be times when it's just the thing to propel you into action.

Allowing yourself to feel your righteous anger about the state of the country your kids are growing up in—no matter what cause may be fueling your passion—makes you more powerful, not less. As a mom, I know you've had to learn how to balance anger and love. And that's a potent recipe for getting back to the work that needs to be done anytime you find your motivation waning. Sadly, there are ample opportunities to feel that anger when there's another shooting; we have to use these tragic incidents to help us stay vigilant so that they happen less and less instead of more and more.

Draw Strength from Camaraderie

Having a strong support system is a crucial component of working through the big emotions that advocacy can stir up. Luckily, getting involved in a cause you care about is also an incredible way to forge deep and meaningful friendships—

the kind of friends who'll be by your side through thick and through thin.

Something I've heard from volunteers over and over is the grateful phrase "I've found my tribe!" It's common to meet people because of an external circumstance you share—your kids are the same age, or you live on the same block, or you were invited to the same party. There's not one thing wrong with that, and I've made many friends this way. But when you connect with others because of a *value* you hold dear, it takes the friendship to another level.

No matter where you live, but especially if you live in a state or community where you feel outnumbered by people who don't share your values, connecting with others who share those views is a balm for the soul. It also provides a support system that helps you stay in the game. You can lean on your soul sisters when the going gets rough, and count on them to pick up the slack when you need a break.

Erica Lafferty Garbatini's mom, Dawn Lafferty Hochsprung, was the Sandy Hook school principal who was murdered as she tried to save the lives of students and educators. Erica started out as a Connecticut Moms Demand Action volunteer and now she's a staff member who says her activist friends feel like a second family. "I lean on the Moms when I need a mom," she says. "They've hugged me, sent me cozy jammies, given me trips to the spa, and opened their homes to me. Through them, I've learned to ask for help before I crash and to accept help when I need it."

Collaborating with others on a cause you care about also helps you stay motivated over the long term, because once you find your tribe, you don't want to leave it. Even if you get tired and take a break, you'll come back as soon as your batteries are recharged not only because you want to keep making progress on the cause, but also because you miss your peeps.

Sara Kerai, a longtime Moms Demand Action volunteer and part of the leadership team of our Washington, DC, chapter, says: "What keeps me going is the utter dedication and professionalism of our volunteers. If they keep going, why would I quit? To watch more TV? This is my tribe now—my Moms sisters have become some of my closest friends."

Having a group of friends with whom you're working toward a common goal also gives you a welcoming space to commiserate after a setback and—even more important—celebrate the successes you've had.

Celebrate Your Wins

You know what's really motivating? Winning. The only problem is that when you're working on a cause that has a lot of fronts and a lot of fires that need to be put out, it's way, way too easy to gloss over your successes and focus on your defeats and setbacks. And when you do that, it's easy to lose sight of all the ground you've covered.

This isn't because of you, personally—it's how our minds are wired. The so-called negativity bias is the evolutionary tendency of our brains to remember everything that goes wrong. It's a survival mechanism, a holdout from the days when we had to remember which berries could kill us or what triggered that bear to attack. It can also make us mentally skip right over anything that doesn't end in disaster.

That's why Moms Demand Action makes it a point to celebrate every win, whether it's an outright victory—such as a passing a good law—or a key defensive play—like killing a bad bill. It's so important to focus on your triumphs, no matter how big or how small, because it reminds you and your tribe that your efforts are making a difference, which, in turn, inspires you to keep going.

At Moms Demand Action, we celebrate in many different ways. Sometimes we throw a big party so volunteers can let loose together. Other times we post a series of social media messages with graphics we've created to help draw more attention to a victory. Or we honor a volunteer or a team with an award at our annual Gun Sense University meeting. I also make it a point to spend a good portion of every speech I give thanking individuals, chapters, and volunteers in general for the incredible gains that they work hard to make. And we have a page devoted to keeping track of our victories on our website (you can check it out for yourself at momsdemandaction.org/our-victories/).

It's also important to publicly celebrate your wins so casual observers are aware of all the little victories that might otherwise go unnoticed. The idea that there hasn't been any progress on gun violence prevention is something I encounter all the time. People often ask me, "How do you keep the faith? Aren't you dejected?" What they don't realize is that we're winning, and winning big. In 2018 alone, we stopped more than one thousand bad gun lobby bills, for a win rate of more than 90 percent for the fourth year in a row. We also helped pass more good gun bills than in any prior legislative session. Even though we make it a point to talk about these successes every chance we get—in press releases, on our social media feeds, in speeches and statements—most people still don't realize how much we've accomplished.

In some ways, I get it: killing bad bills isn't terribly sexy. It doesn't make headlines. Yet it's every bit as important as passing good bills. If Moms Demand Action didn't exist, these bad bills—which seek to pass permitless carry, guns on college campuses, and guns in K-12 schools—would be sailing through statehouses. All our lives would be so much more at-risk and our kids would be even more vulnerable than they already are to school shootings.

For the record, here's an official tally of our wins during our first five years of existence:

- Eleven states have passed laws requiring background checks on all gun sales or strengthening existing background check requirements.
- Twenty-eight states and Washington, DC, have passed laws to help keep guns out of the hands of domestic abusers.
- Eleven states have passed red flag laws to allow a court to temporarily restrict a person's access to guns when that person poses a danger to self or others.
- Eleven states have made bump stocks—gun parts that allow semiautomatic weapons to fire more ammo more rapidly, essentially turning them into a fully automatic machine gun—illegal.
- In 2015, we helped defeat bills allowing guns in K-12 schools in sixteen states, guns on campus in fifteen states, and permitless carry in sixteen states.
- In 2016, we helped defeat bills allowing guns in K-12 schools

in fifteen states, guns on campus in seventeen states, and permitless carry in eighteen states. (As you can see, these bills often come up year after year—all the more reason we need to #KeepGoing.)

- In 2017, we helped defeat bills allowing guns in K-12 schools in twenty-three states, guns on campus in eighteen states, and permitless carry in twenty-six states.
- In 2018, we helped defeat bills allowing guns in K-12 schools in sixteen states, guns on campus in fifteen states, and permitless carry in sixteen states.
- We ended 2013—our first full year in existence—with 80 local groups of volunteers around the country and 125,000 supporters. At the end of 2018, we have 761 local groups and 6 million supporters.
- In 2018, our volunteers made one engagement—such as calling a legislator, hosting a meeting, or staffing a table at an event—for every minute of every day of the entire year.
- Our volunteers educate one new adult on responsible gun practices every fifteen minutes.
- For the first time in 2018, we delivered thousands of Gun Sense Candidate questionnaires to candidates across the country to get politicians on the record on guns. We thought maybe we'd give out a couple hundred designations, but we ended up giving out more than three thousand in forty-eight states.
- In the run-up to the 2018 elections, our volunteers had 1.2 million conversations with voters (that doesn't include all the calls that didn't get answered or doors that didn't get opened) and registered 100,000 people to vote.

- In 2018, forty active Moms Demand Action volunteers ran for office; sixteen of them won.
- Speaking of the 2018 elections, they were a watershed moment for gun safety. We drastically altered the political map by electing gun-sense candidates across the country to offices that range from city council to the US Congress. The legislatures of seven states flipped to become gun-sense majorities—those are seven states where we can now go in and play offense instead of having to stick primarily to defense. We get to play more offense at the federal level too. When I asked Speaker Nancy Pelosi about her legislative agenda in 2019, she told me: "Six years after Sandy Hook, with so much pain and loss between then and now, House Democrats have the opportunity to do what Republican Speakers refused to do, and give the families who experience the agony of losing a loved one to gun violence a vote." And we Moms had a *lot* to do with making the possibility of better federal gun laws so real, by building support for gun violence prevention measures and helping elect a gun-sense majority to the US House of Representatives. Pelosi herself acknowledged our role when she said, "Moms Demand Action has built the public sentiment, the grassroots power, and the local and state precedents to pave the way for federal reform."

There now, doesn't that get you revved up to get out there—whether it's for the first time or the five hundred and first time—and keep showing up?

I hope so, because that's exactly what we each need to do.

Because let's face it, even with all our wins, this kind of activism is never truly done. The NRA was down in the 1990s too, and it came roaring back to life after George W. Bush was elected. There is no "set it and forget it" when it comes to protecting our kids. We have to work to make gains, and then we have to keep working to protect those gains.

The only way it's possible to keep going is to make sure that we're taking care of ourselves and of each other—to find our soul sisters (and brothers) and take one step after the next, together. That's how we save our country from the NRA agenda and how we save lives. It's how we make gun safety our new normal. And it's how we create a bright and safe future for our kids, our grandkids, and our grandkids' kids.

I hope to see you in the fight!

APPENDIX

Talking to Kids About Guns

You may not think you need to teach your kids about guns, but you do. Remember, there are nearly four hundred million guns in circulation in the United States; your kids are likely to be around them no matter how insulated you may think they are. America has a gun culture, and it's up to adults to teach kids how to navigate it, starting from when they are old enough to have conversations up through adulthood.

The Moms Demand Action Be SMART team developed the following tips in collaboration with Marjorie Sanfilippo, a college professor and expert on children's behaviors around firearms.

For young children:

- Make talks about guns and gun safety part of other conversations you have with your kids about topics such as being safe around swimming pools, movies and televisions shows they should avoid, or talking to strangers.
- Keep the language simple; for example: "If you see a gun, don't touch it. Tell an adult right away."
- Tell children not to touch a gun, even if it looks like a toy.
- Assure children that they won't get in trouble if they tell an adult they've seen a gun.
- Repeat your talks regularly.

For older kids:

- Include information about guns and gun safety in your general safety conversations about topics such as drugs, alcohol, and drunk driving.
- Make sure your kids understand that any situation where there's an unsupervised gun is a dangerous situation.
- Tell them to immediately leave any situation—a party, a football game, or any other gathering—where an unsecured gun is present.
- Tell them not to listen to anyone who tells them that a gun is unloaded or otherwise safe.
- Give your teen strategies to get out of a situation where a gun is present—or brainstorm strategies together. For example, you could agree that your teen might say to friends, "Mom just texted me that I have to get home right now."
- Assure them that it's okay to ask about the presence of unsecured guns in other homes, but offer to do it for them if they don't feel comfortable doing it themselves.
- Tell them that if they see or hear something that makes them think a peer might hurt themselves or others with a gun, they should tell a trusted adult. Often, kids who commit school shootings have confided in peers about their plans before they act.
- Keep bringing these issues up frequently, just as you would other crucial safety issues.

As I write this just after the start of the school year, there have already been three incidents across the country where guns caused chaos at high school football games. We can't just hope that our kids won't be exposed to dangerous situations involving guns; just like we need to talk to them about social media, bullying, alcohol, and drugs, we have to talk to them about guns. Doing so might save their lives.

Ensuring Smart Gun Storage in Your Own Home and in Homes Your Kids Will Visit

If a child gets his or her hands on a gun, a bad decision or an honest mistake can easily become fatal. In fact, every year somewhere between two hundred and three hundred children younger than seventeen gain access to a gun and unintentionally shoot themselves or someone else, often fatally.[1] And approximately five hundred children age seventeen or younger kill themselves with a gun each year.[2]

If you own a gun, you may think you have it stored in such a great hiding spot that your children will never find it. But we all know that kids know when there's something you don't want them to see—whether it's the package of your favorite cookies stored on the top shelf of the cupboard or the holiday gifts you've stashed in the back of your closet—and they'll find a way to get at it. The onus should not be on a child or teen to resist picking up a gun they have easy access to or that they find in your hiding place—it's always an adult's responsibility to make sure all guns are stored safely.

That's why we created our Be SMART program, a public health campaign where our volunteers give presentations on how to prevent child suicides by firearms and the unintentional deaths of twenty-five thousand Americans every year.

So, how and where should you keep your guns? They should

be stored unloaded in a locked place (with a gun lock or in a locked location such as a safe, file cabinet, or closet) with the ammunition kept somewhere else.

Your next step is to make sure that any guns in houses where your kids spend time are also stored safely. You may feel confident that those homes don't contain guns, but you could be wrong. Fortunately, all it takes is a simple conversation to keep kids out of harm's way.

I experienced this directly when my son, Sam, was a junior in high school. He was going through a tough time, feeling academically overwhelmed and having issues with his girlfriend; I hadn't seen him that stressed out before. Because my ex-husband Jayson and I share custody, and Sam was due to go stay at Jayson's house, I realized I needed to ask Jayson two things: Did he have any prescription medications that Sam could have easy access to, and where were his guns stored? I didn't think that Sam would harm himself or others, but I was generally worried about him, which brought all kinds of safety concerns to mind. I knew when we were married that Jayson had his dad's hunting rifles stored in the closet with no ammo, but I figured I should ask him about that now. (In retrospect, I see the irony that I didn't think about doing this until five years after starting Moms Demand Action.)

His answer shocked me. He told me he had hunting rifles and a shotgun, with ammo for both, that his dad recently had given him—but they were just in a closet, with the ammo there as well; nothing was secured or locked.

I told Jayson he needed to get gun safes and put the unloaded guns in one safe and the ammo in the other with no way for Sam to access the combination, and I asked him to send me pictures of the secured guns before Sam got there the next day. And he did those things.

You may think you don't need to say anything, or you may feel awkward about asking, but the only way to make sure your kids can't unintentionally get their hands on a gun is to start a conversation. Here are the pointers we share with our volunteers:

- **Make it part of a general safety conversation.** For example, "Before I drop John off for the playdate, I just want to check whether you have a pet, a pool, or firearms in your house. I want to make sure he knows your safety rules."
- **Don't wait to be asked: volunteer information about your own home.** Such as, "We have a pool with a locked gate, and no pets or guns." Or, "Just want you to know we have a cat and a dog, in case Mary has allergies. We also have a hunting rifle, but we always keep it locked and unloaded in a safe."
- **Remember, it's not about the guns; it's about whether they are stored securely.** You don't want to make this an instance when another person could feel judged for having guns. To avoid that, say something like "May I ask whether you have guns in your home, and—if so—are they locked and inaccessible to the kids?"
- **Use technology to your advantage.** If you don't want to talk face-to-face, have the conversation via email or text.
- **Don't forget to talk to family members.** Sixty-five percent of unintentional child deaths by gunfire happen in the homes of relatives. As I experienced with my ex-husband, it's very possible that family members or close friends have unsecured guns in their homes. Many of our volunteers talk

to their own family members as a way to practice bringing the subject up, only to discover that they keep a loaded gun in the spare bedroom in a shoebox. You just can't ever take it for granted. You have to ask.

I hope you'll share this information with friends and family and community members so that we can protect more kids and adults.

Reduce the Risk of Suicide with Smart Gun Storage

A huge part of the reason it's so important to store guns safely is because it reduces the risk of suicide, which is a bigger problem in the United States than most people realize.

The Centers for Disease Control and Prevention estimate that nearly twenty-two thousand Americans die by firearm suicide every year; that translates to about sixty deaths every day.[3] Those numbers are eight times higher than numbers in other high-income countries.[4] They're also on the rise, increasing 19 percent over the past decade.[5]

Guns in homes and suicide go hand in hand: a gun in the house increases the risk of suicide by 300 percent, and this elevated risk applies to everyone in the household—not just the gun owner.[6] Guns are far and away the most lethal method of self-harm, with a fatality rate of nearly 85 percent.[7] Compare that with the fact that fewer than 5 percent of people who attempt suicide using other means die.[8] The sad truth is that guns give a short-term impulse—and nearly half of all survivors of suicide attempts report fewer than ten minutes of deliberation before making the attempt—a final result that affects everyone who loved that person.[9]

Most people who attempt suicide with a gun are adults; adult men represent 86 percent of gun suicide victims.[10] But kids are also uniquely vulnerable to the risk of gun suicide. Nearly one thousand children and teens die by firearm suicide a year, a 61 percent increase over the past decade.[11]

Suicide is an impulsive act, and if you're the parent of a teen, you know how impetuous and emotional they can be. If you or anyone in your family owns a gun, consider these facts:

- 17 percent of American high school students report that they have seriously contemplated committing suicide.[12]
- 4.6 million children and teens in the United States live in a household with at least one loaded, unlocked gun.[13]
- A 2010 study showed that 80 percent of child firearm suicides involved a gun belonging to a family member.[14]

The good news here is that both smart gun storage and commonsense gun laws prevent firearm suicides for people of all ages. Between 1999 and 2013, researchers at Duke University found that police officials in Connecticut (which passed a red flag law in 1999) removed 762 guns from individuals and prevented a suicide for every 10 to 20 guns that were seized.[15] Research has also found that storing guns locked, unloaded, or separate from ammunition reduces the risk of self-inflicted and unintentional firearm injuries among children and teenagers as much as 85 percent, depending on the type of storage practice.[16]

Supporting Yourself, Each Other, and Your Kids After a Mass Shooting

It doesn't matter how old your kids are; mass shootings are upsetting for everyone, of every age. Even my adult children call me when they hear about a shooting tragedy, looking for reassurance that they're safe. Because mass shootings, sadly, have become a fact of American life, Moms Demand Action worked with the National Child Traumatic Stress Network to adapt its guidelines for supporting kids through traumatic events. I include them here:

Begin with yourself. It's important to remember that children's and teens' reactions to shootings are strongly influenced by how parents, relatives, teachers, and other caregivers respond to them. Before you can be of help to others, you need to make sure that you're okay. As in all things related to parenting, you've got to put your own oxygen mask on first so that you're equipped to help others.

Take care of you. Do your best to drink plenty of water, eat regularly, and get enough sleep and exercise. Take time to check in with other adult relatives, friends, or members of the community so that you can support each other. Give yourself permission to limit media input; you don't need to be saturated by 24/7 news. Avoid making any unnecessary life-altering decisions during this time. Take time to rest and do things that you like to do. Doing all these things will help you; they will also model the things that will help your kids feel better too.

Be a positive role model. Share your feelings about the event with your children/teens at a level they can understand. You may express sadness and empathy for the victims and their families. You may share some worry. But it's important to also share ideas for coping with difficult situations and help your children/teens see that there is still good in the world, even in the midst of tragedy, by pointing out the quick responses of law enforcement and medical personnel to help the victims and the heroic or generous efforts of ordinary citizens. Follow the lead of the mother of Fred Rogers (of *Mister Rogers' Neighborhood* fame), who always said to him, "Look for the helpers; you will always find people who are helping" during a crisis.

Start the conversation. Talk about the shooting with your kids. Not talking about it can make the event even more threatening in their minds. Silence suggests that what has occurred is too horrible even to speak about.

Carve out time to have these conversations. Use time such as when you eat together or sit together in the evening to talk about what is happening in the family as well as in the community. Try not to have these conversations close to bedtime, as this is the time for rest and they could lead to nightmares or difficulty falling asleep.

COMMON REACTIONS TO MASS SHOOTINGS

Mass shootings evoke a range of big emotions for you and your kids—you may feel sadness, grief, helplessness, anxiety, and anger. These generally diminish with time, but knowing about them can help you be supportive, of both yourself and your kids. Here are some of the reactions you can expect:

- Anxiety, fear, and worry about the safety of self and others
- Fears that another shooting may occur
- Changes in behavior:
    - Increase in activity level
    - Decrease in concentration and attention
    - Increase in irritability and anger
    - Sadness, grief, and/or withdrawal
    - Radical changes in attitudes and expectations for the future
    - Increases or decreases in sleep and appetite
    - Engaging in harmful habits, like drinking, using drugs, or doing things that are harmful to self or others
    - Lack of interest in usual activities, including spending time with friends
- Physical complaints (headaches, stomachaches, general aches and pains)
- Changes in school and work-related habits and behavior with peers and family

- Staying too focused on the shooting (talking repeatedly about it)
- Strong reactions to reminders of the shooting (media images, smoke, police, memorials)
- Increased sensitivity to sounds (loud noises, screaming)

Find out what your children/teens already know. Start by asking what they have already heard about the events from the media and from friends. As they answer, listen carefully for misinformation, misconceptions, and underlying fears or concerns. Understand that this information will change as more facts about the shooting are known.

Gently correct inaccurate information. If your children/teens have inaccurate information or misconceptions, take time to provide the correct information in simple, clear, age-appropriate language. Like adults, kids are better able to cope with a difficult situation when they have the facts about it.

Encourage your children/teens to ask questions, and answer those questions directly. Having question-and-answer talks gives your kids ongoing support as they begin to cope with the range of emotions stirred up by this tragedy. They may have some difficult questions about the incident. For example, they may ask whether it's possible it could happen to you or them. While it's important to discuss the likelihood of this risk, they are also asking whether they are safe.

Help children/teens feel safe. Talk with them about their concerns over safety and discuss changes that are occurring in the country/community to promote safety. Encourage your children/teens to voice their concerns to you or to teachers at school. This may be a time to review plans your family has for keeping safe in the event of any crisis situation. Also be sure to give any information you have on the help and support the victims and their families are receiving, which gives you an opportunity to highlight the goodness of people.

Encourage self-care. Help kids by encouraging them to drink enough water, eat regularly, and get enough rest and exercise. Let them know it's okay to take a break from talking with others about the recent attack or from participating in any of the memorial events.

Maintain expectations or "rules." Stick with family rules, such as curfews, checking in, and keeping up with homework and chores. On a time-limited basis, keep a closer watch on where teens are going and what they are planning to do in order to monitor how they are doing. Assure them that the extra check-in is temporary—just until things stabilize.

Expect mood changes. Children/teens may become more irritable or defiant. They may have trouble separating from caregivers, wanting to stay at home or close by them. It's common for young people to feel anxious about what has happened, what may happen in the future, and how it will impact their lives. Remember, they may be thinking about the event even when they try not to. Their sleep and appetite routines may change. In general, you should see these reactions lessen within a few weeks.

Limit media exposure. Especially limit your younger children's exposure to media images and sounds of the shooting, and do not allow your very young children to see or hear any media information about the shootings. Even if they appear to be engrossed in play, children often are aware of what you are watching on TV or listening to on other media. What may not be upsetting to an adult may be very upsetting and confusing for a child. Limit your own exposure as well. Adults may become more distressed with nonstop exposure to media coverage of shootings.

Be patient. In times of stress, children/teens may have trouble with their behavior, concentration, and attention. They may not openly ask for your guidance or support, but they will want it. Adolescents who are seeking increased independence may have difficulty expressing their needs. Both children and teens will need a little extra patience, care, and love. (Be patient with yourself, too!) Accept responsibility for your own feelings, by saying things like "I want to apologize for being irritable with you yesterday. I was having a bad day."

Monitor changes in relationships. Explain to kids that strains on relationships are to be expected. Emphasize that everyone needs family and friends for support during this time. Spend more time talking as a family about how everyone is doing. Encourage tolerance for how your family and friends may be recovering or feeling differently.

Address radical changes in attitudes toward and expectations for the future. Explain to children/teens that changes in people's attitudes are common and tend to be temporary after a tragedy like this. These feelings can include feeling scared, angry, and sometimes revengeful. Find other ways to make them feel more in control and talk about their feelings.

Identify constructive activities. Children and teens are often deeply concerned for survivors and families of survivors and want to help them. Encourage age-appropriate activities that are meaningful (collecting money, supplies, etc.) as long as it seems helpful and not burdensome.

Get extra help when you need it. Should reactions continue or at any point interfere with your children's/teens' abilities to function, or if you are worried, contact local mental health professionals who have expertise in trauma. Contact your family physician, pediatrician, or state mental health associations for referrals to such experts.

As helpful as these guidelines can be, the truth is that they are all Band-Aids. It's hard to assure kids of any age that they're safe and that a mass shooting can't happen where you live; I don't know whether any American parent really believes that anymore. Unless you also get out there and fight for stronger gun laws, there won't be a change. Everything I've shared in these appendix resources are just more reasons to get involved. We, your soul sisters, are ready and excited to welcome you into the fold.

Tips compiled and adapted by Ava Schlesinger, LCSW, from sup-

port resources published by the National Child Traumatic Stress Network, http://www.nctsn.org.

Public Distress Helpline

If you or anyone you know needs emotional support as a result of gun violence, please urge them to contact the Substance Abuse and Mental Health Services Administration (SAMHSA) Distress Helpline, which offers crisis support services for any American experiencing emotional distress, at 1-800-985-5990 or text "TalkWithUs" to 66746.

ACKNOWLEDGMENTS

This book has been a labor of love—a fitting description for a book about how to give birth to a movement and fight like a mother. And, of course, it was created through collaboration among so many "mothers and others" across the nation.

First and foremost, I dedicate this book to my children for making me a mom in the first place: Abigail, Emma, Samuel, Kelly, and Samantha. My concern for your safety has fueled me to fight this fight every day since 2012. And to my kids' dad, Jayson: thank you for consciously uncoupling with me before it was cool.

To my husband, John, the love of my life and a constant source of wisdom, kindness, and support: I walk around bearing the blissful secret that I won the life-partner lottery. Thank you for loving me unconditionally.

To every Moms Demand Action volunteer and gun violence survivor who has or will give his or her time, talent, and passion to our organization: this book is because of you and it is in honor of you. I was naive enough to start a Facebook page and brash enough to serve as the tip of the spear since Moms Demand Action's inception, but you are its heart and soul. This organization will last into perpetuity because you show up day after day and selflessly do the unglamorous heavy lifting of grassroots activism.

To the volunteers and experts who helped me write this book, including Kara Waite and Kate Hanley, thank you for your patience and time. Kate, thank you for making these stories sing.

Thank you to my editor Hilary Swanson, and the wonderful team at HarperOne: Laina Adler, Suzanne Wickham, Melinda Mullin, Julia Kent, Suzanne Quist, and Adrian Morgan.

I also want to thank my agent, Kathy Schneider, and Chris Prestia and the entire Jane Rotrosen Agency for their support.

And to everyone at Everytown and Moms Demand Action for your support and dedication. In particular, John Feinblatt, Amanda Konstam, Taylor Maxwell, Aimee Tavares, Meghan Adamoli, and Tracy Sefl for all of their help throughout the process.

Finally, there are so many more stories I wish I could have told, and so many more volunteers I wish I'd had room to shout out to, but a book can only have so many pages. I hope every volunteer will see themselves in the pages of *Fight Like a Mother*, and know deep down that every effort they've made and every action they've taken has helped build one of the largest grassroots movements in the nation. Keep going . . .

NOTES

INTRODUCTION

1. A. Karp, "Estimating Global Civilian-Held Firearms Numbers," Small Arms Survey, June 2018, http://www.smallarmssurvey.org/fileadmin/docs/T-Briefing-Papers/SAS-BP-Civilian-Firearms-Numbers.pdf.
2. US Census Bureau, "U.S. and World Population Clock," https://www.census.gov/popclock/, accessed October 31, 2018.
3. Karp, "Estimating."
4. K. D. Kochanek, S. L. Murphy, J. Xu, and B. Tejada-Vera, "Deaths: Final Data for 2014," *National Vital Statistics Reports* 65, no. 4 (2016), https://www.cdc.gov/nchs/data/nvsr/nvsr65/nvsr65_04.pdf.
5. S. L. Murphy, J. Xu, K. D. Kochanek, S. C. Curtin, and E. Arias, "Deaths: Final Data for 2015," *National Vital Statistics Reports* 66, no. 6 (2017), https://www.cdc.gov/nchs/data/nvsr/nvsr66/nvsr66_06.pdf.
6. J. Xu, S. L. Murphy, K. D. Kochanek, B. Bastian, and E. Arias, "Deaths: Final Data for 2016," *National Vital Statistics Reports* 67, no. 5 (2018), https://www.cdc.gov/nchs/data/nvsr/nvsr66/nvsr66_06.pdf.
7. Centers for Disease Control and Prevention, WONDER (Wide-Ranging Online Data for Epidemiologic Research), "Underlying Cause of Death," https://wonder.cdc.gov/ucd-icd10.html. Data reflect a five-year average (2013–2017) of gun deaths by intent.
8. M. Fox, "Guns Send 8300 Kids to Hospital Every Year, Study Finds," NBC News, October 28, 2018, https://www.nbcnews.com/health/health-news/guns-send-8-300-kids-hospitals-each-year-study-finds-n925906?cid=sm_npd_nn_tw_ma.
9. Centers for Disease Control and Prevention, Injury Prevention and Control, WISQARS (Web-Based Injury Statistics Query and Reporting System), "Fatal Injury Data," https://www.cdc.gov/injury/wisqars/index.html. Data reflect a five-year average (2012–2016) of gun deaths by race. Analysis includes all ages, non-Hispanic only, and homicide, including legal intervention.
10. Ibid. Data reflect a five-year average (2012–2016) of gun deaths by race. Analysis includes ages zero to nineteen, and non-Hispanic only and homicide, including legal intervention.
11. Department of Justice, Federal Bureau of Investigation, "Uniform Crime Reporting Program: Supplementary Homicide Report (SHR), 2012–2016," Washington, DC. While the FBI SHR does not include data from the state of Florida for the years 2012–2016, Everytown for Gun Safety obtained data directly from the Florida Department of Law Enforcement (FDLE) and included the reported homicides in the analysis. Whereas the SHR includes both current and former partners in its relationship designations, the FDLE does not include former partners. As a result, Florida's intimate partner violence data includes only current partners.
12. E. Grinshteyn and D. Hemenway, "Violent Death Rates: The U.S.

Compared with Other High-Income OECD Countries, 2010," *American Journal of Medicine* 129, no. 3 (2016): 266–273.

13. Global Burden of Disease 2016 Injury Collaborators, "Global Mortality from Firearms, 1990–2016," *JAMA* 320, no. 8 (2018): 792–814.
14. Worldometers, "U.S. Population (Live)," http://www.worldometers.info/world-population/us-population/, accessed November 30, 2018.

CHAPTER 1: Use MOMentum

1. National Conference of State Legislatures, "Women in State Legislatures for 2019," January 8, 2019, http://www.ncsl.org/legislators-staff/legislators/womens-legislative-network/women-in-state-legislatures-for-2019.aspx.
2. D. Desilver, "A Record Number of Women Will Be Serving in the New Congress," *FactTank*, Pew Charitable Research, December 18, 2018, http://www.pewresearch.org/fact-tank/2018/12/18/record-number-women-in-congress.
3. J. Cooperman, "The Making of the NRA's Dana Loesch," *St. Louis Magazine*, September 13, 2018, https://www.stlmag.com/longform/the-making-of-dana-loesch/.

CHAPTER 3: Channel Your Inner Badass

1. M. Blumenthal, "Palin's New Disaster," *The Daily Beast*, April 13, 2009, https://www.thedailybeast.com/palins-new-disaster.
2. T. Johnston, "10 Misogynist Attacks from Ted Nugent, Greg Abbott's New Surrogate," Media Matters for America, February 14, 2014, https://www.mediamatters.org/research/2014/02/14/10-misogynist-attacks-from-ted-nugent-greg-abbo/198061.
3. NRA-ILA, Institute for Legislative Action, "Rhode Island: House Judiciary Committee Preparing for Gun Control Round Two," May 8, 2017, https://www.nraila.org/articles/20170508/rhode-island-house-judiciary-committee-preparing-for-gun-control-round-two.
4. T. Johnson, "The NRA Is Trying to Reach Out to Women, but This Is How It Talks About Campus Sexual Assault," Media Matters for America, September 9, 2014, https://www.mediamatters.org/blog/2014/09/09/the-nra-is-trying-to-reach-out-to-women-but-thi/200699.

CHAPTER 4: Losing Forward

1. C. Cheng and M. Hoekstra, "Does Strengthening Self-Defense Law Deter Crime or Escalate Violence? Evidence from Castle Doctrine," *Journal of Human Resources* 48, no. 3 (2013): 821–854, http://econweb.tamu.edu/mhoekstra/castle_doctrine.pdf.

CHAPTER 5: Use Your Bullhorn

1. J. Adalian, "Late-Night Ratings: Stephen Colbert's Lead Over Jimmy Fallon Is Bigger than Ever," Vulture, April 16, 2018, https://www.vulture.com/2018/04/late-night-ratings-early-2018-colbert-fallon.html.

2. K. Bialik, "14% of Americans Have Changed Their Mind About an Issue Because of Something They Saw on Social Media," Pew Research Center, August 15, 2018, http://www.pewresearch.org/fact-tank/2018/08/15/14-of-americans-have-changed-their-mind-about-an-issue-because-of-something-they-saw-on-social-media/.
3. City of New York, "Point, Click, Fire: An Investigation of Illegal Online Gun Sales," December, 2011, http://everytownresearch.org/documents/2015/04/point-click-fire.pdf; K. A. Vittes, J. S. Vernic, and D. W. Webster, "Legal Status and Source of Offenders' Firearms in States with the Least Stringent Criteria for Gun Ownership," *Injury Prevention* 19 (2013): 26–31.

CHAPTER 6: Tap into the Priceless Power of Volunteers

1. Quinnipiac University, "U.S. Voters Reject GOP Health Plan More than 3–1, Quinnipiac University National Poll Finds; Voters Support Gun Background Checks 94–5 Percent," June 28, 2017, https://poll.qu.edu/images/polling/us/us06282017_Uvx74gpk.pdf/.

CHAPTER 8: Know Your Numbers

1. A. Milkovits, "In R.I. Domestic-Violence Cases, Suspects Often Keep Guns," *Providence Journal*, June 16, 2015, https://www.providencejournal.com/article/20150616/NEWS/150619505.
2. A. Kellermann, F. P. Rivara, N. B. Rushforth, J. G. Banton, D. T. Reay, J. T. Francisco, A. B. Locci, J. Prodzinski, B. B. Hackman, and G. Somes, "Gun Ownership as a Risk Factor for Homicide in the Home," *New England Journal of Medicine* 329 (1993): 1084–1091.
3. A. Kellermann and F. P. Rivara, "Silencing the Science on Gun Research," *JAMA* 309, no. 6 (2013): 549–550.
4. Omnibus Consolidated Appropriations Bill, HR 3610, Pub. L. No. 104-208, September 1996, http://www.gpo.gov/fdsys/pkg/PLAW-104publ208/pdf/PLAW-104publ208.pdf.
5. J. R. Lott and D. B. Mustard, "Crime, Deterrence, and Right-to-Carry Concealed Weapons," *Journal of Legal Studies* 26, no. 1 (1997): 1–68.
6. C. F. Wellford, J. V. Pepper, and C. V. Petrie, *Firearms and Violence: A Critical Review* (Washington, DC: National Academies Press, 2005); J. J. Donohue, A. Aneja, and A. Zhang, "The Impact of Right to Carry Laws and the NRC Report: The Latest Lessons for the Empirical Evaluation of Law and Policy," Stanford Law and Economics Olin Working Paper No. 430, July 29, 2012.
7. J. Lurie, "When the Gun Lobby Tries to Justify Fire Arms Everywhere, It Turns to This Guy," *Mother Jones*, July 28, 2015, https://www.motherjones.com/politics/2015/07/john-lott-guns-crime-data/.
8. J. R. Lott, "Why You Should Be Hot and Bothered About 'Climate-Gate,'" FoxNews, November 24, 2009, last updated May 7, 2015, https://www.foxnews.com/opinion/why-you-should-be-hot-and-bothered-about-climate-gate; O. Exstrum, "The Guy Behind the Bogus Immigration Report Has a Long History of Terrible and Misleading Research," *Mother Jones*, February 7, 2018, https://www.motherjones.com/politics/2018/02

/the-guy-behind-the-bogus-immigration-report-has-a-long-history-of-terrible-and-misleading-research/.

9. A. Moodie, "Before You Read Another Health Study, Check Who's Funding the Research," *Guardian,* December 12, 2016; A. O'Connor, "Coca-Cola Funds Scientists Who Shift Blame for Obesity Away from Bad Diets," NYTimes.com, August 9, 2015, https://www.theguardian.com/lifeandstyle/2016/dec/12/studies-health-nutrition-sugar-coca-cola-marion-nestle.
10. A. Karp, "Estimating Global Civilian-Held Firearms Numbers," Small Arms Survey, June 2018, http://www.smallarmssurvey.org/fileadmin/docs/T-Briefing-Papers/SAS-BP-Civilian-Firearms-Numbers.pdf.
11. These and following statistics from E. Grinshteyn and D. Hemenway, "Violent Death Rates: The US Compared with Other High-Income OECD Countries, 2010," *American Journal of Medicine* 129, no. 3 (2016): 266–273.
12. Sam Stein, "Gun Owners Surveyed by Frank Luntz Express Broad Support for Gun Control Policies," *Huffington Post,* July 24, 2012, https://www.huffingtonpost.com/2012/07/24/gun-owners-frank-luntz-n-1699140.html.
13. City of Chicago, Office of the Mayor, "Gun Trace Report, 2017," https://www.chicago.gov/content/dam/city/depts/mayor/Press%20Room/Press%20Releases/2017/October/GTR2017.pdf.
14. J. C. Karberg, R. J. Frandsen, J. M. Durst, T. D. D. Buskirk, and A. D. Lee, "Background Checks for Firearm Transfers, 2013–14: Statistical Tables," US Department of Justice, Bureau of Justice Statistics, June 2016, http://bit.ly/2av5tvL. Data for 2015 and 2016 were obtained by Everytown for Gun Safety from the FBI directly. Though the majority of transactions and denials reported by the FBI and Bureau of Justice Statistics are associated with a firearm sale or transfer, a small number may be for concealed-carry permits and other reasons not related to a sale or transfer.
15. K. E. Rudolph, E. A. Stuart, J. S. Vernick, and D. W. Webster, "Association Between Connecticut's Permit-to-Purchase Handgun Law and Homicides," *American Journal of Public Health* 105, no. 8 (2015): e49–e54.
16. D. Webster, C. K. Crifasi, and J. S. Vernic, "Effects of Missouri's Repeal of Its Handgun Purchaser Licensing Law on Homicides," Johns Hopkins Bloomberg School of Public Health, December 17, 2013, https://www.jhsph.edu/research/centers-and-institutes/johns-hopkins-center-for-gun-policy-and-research/_pdfs/effects-of-missouris-repeal-of-its-handgun-purchaser-licensing-law-on-homicides.pdf.
17. A. Baker, "A Hail of Bullets, a Heap of Uncertainty," *New York Times,* December 9, 2007.
18. J. Hanna and H. Yan, "Sutherland Springs Church Shooting: What We Know," CNN, November 7, 2017, https://www.cnn.com/2017/11/05/us/texas-church-shooting-what-we-know/index.html; R. Bustamente, "Autopsy for Texas Church Gunman Confirms Death by Suicide," *USA Today,* August 29, 2018.
19. W. J. Krouse and D. J. Richardson, "Mass Murder with Firearms: Incidents and Victims, 1999–2013," Congressional Research Service, July 30, 2015, https://fas.org/sgp/crs/misc/R44126.pdf.
20. "Public Law 112-265—January 14, 2013," Authenticated U.S. Government

Information, GPO, https://www.govinfo.gov/content/pkg/PLAW-112publ265/pdf/PLAW-112publ265.pdf.

21. Everytown for Gun Safety, "Mass Shootings in the United States: 2009–2016," March 2017, https://everytownresearch.org/wp-content/uploads/2017/03/Analysis_of_Mass_Shooting_033117.pdf.
22. World Health Organization, *Depression and Other Common Mental Disorders: Global Health Estimates* (Geneva: World Health Organization, 2017), 8 and 10.
23. J. Y. Choe, L. A. Teplin, and K. M. Abram, "Perpetration of Violence, Violent Victimization, and Severe Mental Illness: Balancing Public Health Concerns," *Psychiatric Services* 59, no. 2 (2008): 153–164.
24. Everytown for Gun Safety, "Mass Shootings," 2.
25. M. Livingston, "More States Approving 'Red Flag' Laws to Keep Guns Away from People Perceived as Threats," *Los Angeles Times,* May 14, 2018.
26. Cornell University Law School, Legal Information Institute, *District of Columbia et al. v. Heller* (No. 07-290), 478 F. 3d 370, argued March 18, 2008, decided June 26, 2008, https://www.law.cornell.edu/supct/html/07-290.ZS.html.
27. D. Hemenway and E. G. Richardson, "Homicide, Suicide, and Unintentional Firearm Fatality: Comparing the United States with Other High-Income Countries, 2003," *Journal of Trauma* 70 (2011): 238–242.
28. Five times more likely: J. C. Campbell, S. W. Webster, J. Koziol-McLain, et al., "Risk Factors for Femicide in Abusive Relationships: Results from a Multisite Case Control Study," *American Journal of Public Health* 93 (2003): 1089–1097; fifty American women: Federal Bureau of Investigation, "Supplementary Homicide Reports, 2009–13," https://www.icpsr.umich.edu/icpsrweb/NACJD/series/57/studies#; one million American women and four and a half million American women: S. B. Sorenson and R. A. Schut, "Nonfatal Gun Use in Intimate Partner Violence: A Systematic Review of the Literature," *Trauma, Violence, and Abuse* 19, no. 4 (2016): 431–442.
29. Between the inception of the NICS system in 1998 and December 31, 2018, 145,826 gun sales were federally denied due to a misdemeanor crime of domestic violence conviction, and 60,254 gun sales were federally denied due to restraining or protection orders for domestic violence, making a total of 206,080 federal denials related to domestic violence. US Department of Justice, Federal Bureau of Investigation, "NICS Denials: Reasons Why the NICS Section Denies, Nov. 1, 1998–Dec. 31, 2018." Between 1998 and 2010, state and local agencies issued a total of 945,915 denials, and for agencies that reported reasons for these denials, 13.2 percent were denials for domestic violence reasons—which would represent another 124,861 domestic violence denials. R. J. Frandsen, D. Naglich, G. A. Lauver, and A. D. Lee, "Background Checks for Firearms Transfers, 2010: Statistical Tables," US Department of Justice, Bureau of Justice Statistics, February 2013, http://1.usa.gov/Z8vYsa. Between 2012 and 2014, state and local agencies reportedly issued an additional 18,578 domestic violence–related denials. J. C. Karberg, R. J. Frandsen, J. M. Durso, and A. D. Lee, "Background Checks for Firearms Transfers, 2012: Statistical Tables," US Department of Justice, Bureau of Justice Statistics,

December 9, 2014, http://www.bjs.gov/index.cfm?ty=pbdetail&iid=5157; J. C. Karberg, R. J. Frandsen, J. M. Durso, T. D. Buskirk, and A. D. Lee, "Background Checks for Firearms Transfers, 2013–2014: Statistical Tables," US Department of Justice, Bureau of Justice Statistics, June 30, 2016, http://www.bjs.gov/index.cfm?ty=pbdetail&iid=5664. Thus, overall the background check system has issued an estimated 304,234 denials due to domestic violence–related criteria between 1998 and 2014. This is likely to be an underestimation since it does not include state and local denials data for 2011 and local denials data for 2013.

30. US Department of Justice, Federal Bureau of Investigation, "NICS Operations Reports, 1998–2013," https://www.fbi.gov/resources/library.
31. US Department of Justice, Federal Bureau of Investigation, "Uniform Crime Reporting Program Data: Supplementary Homicide Reports, 2011 (ICPSR 34588)," https://doi.org/10.3886/ICPSR34588.v1, excludes New York because of incomplete data; Florida Department of Law Enforcement, "Supplemental Homicide Report, 2010," http://www.fdle.state.fl.us/FSAC/Crime-Data/SHR.aspx.
32. Mayors Against Illegal Guns, "Felon Seeks Firearm, No Strings Attached: How Dangerous People Evade Background Checks and Buy Illegal Guns Online," September 2013, http://bit.ly/1nllhRb.
33. A. Cooper and E. L. Smith, "Homicide Trends in the United States, 1980–2008," US Department of Justice, November 2011, http://1.usa.gov/1fpGIbN.
34. Only fifteen states prohibit all domestic violence misdemeanants and subjects of restraining orders from buying or owning guns: California, Colorado, Connecticut, Delaware, Hawaii, Illinois, Iowa, Louisiana, Minnesota, New Jersey, New York, Tennessee, Texas, Washington, and West Virginia.
35. "American Outdoor Brands Corporation Reports Third Quarter Fiscal 2018 Financial Results," American Outdoor Brands, March 1, 2018, http://ir.smith-wesson.com/phoenix.zhtml?c=90977&p=RssLanding&cat=news&id=2335745; M. Haag, "Remington, Centuries-Old Gun Maker, Files for Bankruptcy as Sales Slow," *New York Times,* March 25, 2018.
36. K. Washburn and R. Maguire, "Member Dues Plummet, Leaving the NRA in the Red for Second Straight Year," OpenSecrets.org, Center for Responsive Politics, September 19, 2018, https://www.opensecrets.org/news/2018/09/nra-in-the-red-for-2nd-straight-year/.
37. "DFS Fines Lockton Companies $7 Million for Underwriting NRA-Branded 'Carry Guard' Insurance Program in Violation of New York Insurance Law," New York State Department of Financial Services press release, May 2, 2018, https://www.dfs.ny.gov/about/press/pr1805021.htm.
38. M. T. Vulla, "Guidance on Risk Management Relating to the NRA and Similar Gun Promotion Organizations," New York State Department of Financial Services memorandum, April 19, 2018, https://www.dfs.ny.gov/legal/dfs/DFS_Guidance_Risk_Management_NRA_Gun_Manufacturers-Insurance.pdf.
39. T. Dickinson, "The NRA Says It's in Deep Financial Trouble, May Be 'Unable to Exist,'" *Rolling Stone,* August 3, 2018, https://www.rollingstone.com/politics/politics-news/nra-financial-trouble-706371/.

40. Associated Press, "These Are the Companies That Have Cut Ties with the NRA," *Los Angeles Times,* May 2, 2018.
41. "National Rifle Association: Outside Spending Summary, 2012," OpenSecrets.org, Center for Responsive Politics, https://www.opensecrets.org/outsidespending/detail.php?cmte=National+Rifle+Assn&cycle=2012.
42. Centers for Disease Control and Prevention, Injury Prevention and Control, WISQARS (Web-Based Injury Statistics Query and Reporting System), "Fatal Injury Data," https://www.cdc.gov/injury/wisqars/index.html. Data reflect a five-year average (2012–2016) of gun deaths by race. Analysis includes all ages, non-Hispanic only, and homicide, including legal intervention.

CHAPTER 9: Build a Big Tent

1. Centers for Disease Control and Prevention, Injury Prevention and Control, WISQARS (Web-Based Injury Statistics Query and Reporting System), "Nonfatal Injury Data," https://www.cdc.gov/injury/wisqars/index.html. Data reflect a five-year average (2012–2016) of gun deaths by intent. Analysis includes males of all ages, non-Hispanic only, and homicide, including legal intervention.
2. Ibid., "Fatal Injury Data." Data reflect a five-year average (2012–2016) of gun deaths by race. Analysis includes ages zero to nineteen, non-Hispanic only, and homicide, including legal intervention.
3. Department of Justice, Federal Bureau of Investigation, "Uniform Crime Reporting Program: Supplementary Homicide Reports (SHR), 2012–2016," https://www.icpsr.umich.edu/icpsrweb/NACJD/series/57/studies#. Analysis by Everytown Research includes homicides involving an intimate partner and a firearm and compares the crude death rates for black women (0.63 per 100,000) versus white women (0.34 per 100,000) (all ages included; Hispanic and non-Hispanic women included).
4. Mapping Police Violence, https://mappingpoliceviolence.org/, last updated December 21, 2018.
5. M. Spies, "The N.R.A. Lobbyist Behind Florida's Pro-Gun Policies," *New Yorker,* March 5, 2018.
6. N. Ackermann, M. S. Goodman, K. Gilbert, C. Arroyo-Johnson, and M. Pagano, "Race, Law, and Health: Examination of 'Stand Your Ground' and Defendant Convictions in Florida," *Social Science and Medicine* 142 (2015): 194–201.
7. J. K. Roman, "Race, Justifiable Homicide, and Stand Your Ground Laws: Analysis of FBI Supplementary Homicide Report Data," Urban Institute, July 2013, https://www.urban.org/sites/default/files/publication/23856/412873-Race-Justifiable-Homicide-and-Stand-Your-Ground-Laws.pdf.
8. E. Bazelon, "What If Trayvon Martin Was the One Acting in Self-Defense?," Slate, March 22, 2012, http://www.slate.com/articles/news_and_politics/crime/2012/03/floridas_stand_your_ground_law_doesnt_prohibit_that_they_arrest_george_zimmerman_for_killing_trayvon_martin.html.
9. NCSL, National Conference of State Legislators, "Self Defense and 'Stand

Your Ground,'" July 27, 2018, http://www.ncsl.org/research/civil-and-criminal-justice/self-defense-and-stand-your-ground.aspx.

10. J. Soltz, "George Zimmerman Had More Legal Authority to Kill Than Our Troops Do at War," Think Progress, April 10, 2012, http://thinkprogress.org/justice/2012/04/10/460965/zimmerman-shoot-kill-troops-military/.
11. C. Cheng and M. Hoekstra, "Does Strengthening Self-Defense Law Deter Crime or Escalate Violence? Evidence from Castle Doctrine," *Journal of Human Resources* 48, no. 3 (2013): 821–854, http://econweb.tamu.edu/mhoekstra/castle_doctrine.pdf.
12. M. Masucci and L. Langton, "Special Report: Hate Crime Victimization, 2004–2015," US Department of Justice, Bureau of Justice Statistics, June 2017, https://www.bjs.gov/content/pub/pdf/hcv0415.pdf. To obtain the annual and daily average of hate crimes involving a firearm, Everytown for Gun Safety used a ten-year average of violent hate crime victimizations (2006–2015) combined with the proportion of violent hate crimes involving firearms (4.5 percent). Analysis was limited to violent hate crimes perpetrated against a person or people and does not include hate crimes against property (such as defacing a victim's home, burglary, and vehicle theft).
13. US Department of Justice, Federal Bureau of Investigation, "Hate Crime Statistics, 2016," November 2017, https://ucr.fbi.gov/hate-crime/2016/topic-pages/incidentsandoffenses. Everytown notes on its website: "It is important to note that the FBI UCR data on hate crimes is a severe undercount since most participating law enforcement agencies do not report these data to the FBI. The FBI UCR data were used to understand bias motivations since a breakdown for single bias incidents is provided unlike the NCVS data, which for all other purposes, is a more complete source of data for hate crime victimizations" (https://everytownresearch.org/disarm-hate/#foot_note_6).
14. Southern Poverty Law Center, "Hate Groups Increase for Second Consecutive Year as Trump Electrifies Radical Right," February 15, 2017, https://www.splcenter.org/news/2017/02/15/hate-groups-increase-second-consecutive-year-trump-electrifies-radical-right; R. Cohen, "Hate Crimes Rise for Second Straight Year; Anti-Muslim Violence Soars Amid President Trump's Xenophobic Rhetoric," Southern Poverty Law Center, November 13, 2017, https://www.splcenter.org/news/2017/11/13/hate-crimes-rise-second-straight-year-anti-muslim-violence-soars-amid-president-trumps; E. Abdelkader, "When Islamophobia Turns Violent: The 2016 U.S. Presidential Elections," Georgetown University, posted May 15, 2016, https://ssrn.com/abstract=2779201.
15. Anti-Defamation League, "U.S. Anti-Semitic Incidents Spike 86 Percent So Far in 2017 After Surging Last Year, ADL Finds," April 24, 2017, https://www.adl.org/news/press-releases/us-anti-semitic-incidents-spike-86-percent-so-far-in-2017.
16. E. Waters, L. Pham, C. Convery, and S. Yacka-Bible, "A Crisis of Hate: A Report on Lesbian, Gay, Bisexual, Transgender and Queer Hate Violence Homicides in 2017," National Coalition of Anti-Violence Programs, https://avp.org/wp-content/uploads/2018/01/a-crisis-of-hate-january-release.pdf.

17. Human Rights Campaign, "Violence Against the Transgender Community in 2018," https://www.hrc.org/resources/violence-against-the-transgender-community-in-2018.
18. A. Aufrichtig, L. Beckett, J. Diehm, and J. Lartey, "Want to Fix Gun Violence in America? Go Local," *Guardian*, January 9, 2017.
19. R. J. Epstein, "The Democratic Party's New Litmus Test: Gun Control," *Wall Street Journal*, August 9, 2018.
20. "'My Third Conversion': Rev. Rob Schenck on Why He Took on Gun Control," *All Things Considered*, May 27, 2018, https://www.npr.org/2018/05/27/614886515/-my-third-conversion-rev-rob-schenck-on-why-he-took-on-gun-control.

CHAPTER 10: Let This Mother Run This Mother

1. Rutgers Center for American Women and Politics, "Women in Elective Office, 2019," accessed January 25, 2019, http://www.cawp.rutgers.edu/women-elective-office-2019.
2. Center for Public Integrity, "Women Will Hold Record Numbers of Elected Offices in 2019. See Where They Made the Biggest Gains," December 19, 2018, https://publicintegrity.org/state-politics/share-of-women-in-elected-office-in-every-state/.
3. Rutgers Center for American Women and Politics, "Current Numbers," accessed January 25, 2019, http://www.cawp.rutgers.edu/current-numbers.
4. K. Ziegler, "Female Candidates Win in Historic Numbers," National Conference of State Legislators, November 8, 2018, http://www.ncsl.org/blog/2018/11/08/female-candidates-win-in-historic-numbers.aspx.
5. K. Sanbonmatsu, S. J. Carroll, and D. Walsh, "Poised to Run: Women's Pathways to the State Legislatures," Rutgers Center for American Women and Politics (New Jersey: Rutgers, the State University of New Jersey, 2009), http://www.cawp.rutgers.edu/sites/default/files/resources/poisedtorun_0.pdf.
6. J. L. Lawless and R. L. Fox, "Girls Just Wanna Not Run: The Gender Gap in Young Americans' Political Ambition," American University School of Public Affairs, 2013, https://www.american.edu/spa/wpi/upload/girls-just-wanna-not-run_policy-report.pdf.
7. K. F. Kahn and E. N. Goldenberg, "Women Candidates in the News: An Examination of Gender Differences in U.S. Senate Campaign Coverage," *Public Opinion Quarterly* 55, no. 2 (1991): 180–199; M. C. Bligh, M. M. Schlehofer, B. J. Casad, and A. M. Gaffney, "Competent Enough, But Would You *Vote* for Her? Gender Stereotypes and Media Influences on Perceptions of Women Politicians," *Journal of Applied Social Psychology* 42, no. 3 (2012): 560–597.
8. C. H. Mo, "The Consequences of Explicit and Implicit Gender Attitudes and Candidate Quality in the Calculations of Voters," *Political Behavior* 37 (2012): 357.
9. J. Lazarus and A. Steigerwalt, *Gendered Vulnerability: How Women Work Harder to Stay in Office* (Michigan: Michigan Univ. Press, 2018), https://www.press.umich.edu/9718645/gendered_vulnerability.

10. C. C. Miller, "Women Actually Do Govern Differently," *New York Times*, November 10, 2016, https://www.nytimes.com/2016/11/10/upshot/women-actually-do-govern-differently.html.
11. Center for American Women and Politics, Rutgers Eagleton Institute of Politics, "What Role Will Women Play in the Legislative Debate Over Gun Control?," http://www.cawp.rutgers.edu/footnotes/what-role-will-women-play-legislative-debate-over-gun-control, accessed December 28, 2018.
12. M. Skoneki, "In Tallahassee, It's Almost Always Marion Hammer Time, Emails Show," *Orlando Sentinel*, September 23, 2018.
13. Quinnipiac University, "Americans Have Little Hope for World Peace in 2018, Quinnipiac University National Poll Finds; 'Merry Christmas' Is Bogus Issue, Voters Say 4–1," December 20, 2017, https://poll.qu.edu/images/polling/us/us12202017_uvc5698.pdf/.
14. Rutgers Center for American Women and Politics, "Women in the U.S. Congress 2019," http://www.cawp.rutgers.edu/women-us-congress-2019, accessed January 25, 2019.
15. Allianz Life Insurance Company of North America, "The Allianz Women, Power, and Money Study: Empowered and Underserved," October 2016, https://www.allianzlife.com/-/media/files/allianz/documents/ent_1462_n.pdf?la=en&hash=DB76F6EE3B711B77523AABC237F9B37F6E8F2F21.
16. US Department of Agriculture, Center for Nutrition Policy and Promotion, "Expenditures on Children by Families Report, 2015," miscellaneous report no. 1528-2015, January 2017, revised March 2017, https://www.cnpp.usda.gov/sites/default/files/crc2015.pdf.

APPENDIX: Talking to Kids About Guns

1. Everytown for Gun Safety, "#NotAnAccident Index," https://everytownresearch.org/notanaccident/, accessed January 25, 2019.
2. K. A. Fowler, L. L. Dahlberg, T. Haileyesus, C. Gutierrez, and S. Bacon, "Childhood Firearm Injuries in the United States," *Pediatrics* 140, no. 1 (2017), http://pediatrics.aappublications.org/content/140/1/e20163486.
3. Centers for Disease Control and Prevention, Injury Prevention and Control, WISQARS (Web-Based Injury Statistics Query and Reporting System), https://www.cdc.gov/injury/wisqars/index.html. Averages developed using five years of data, 2012–2016.
4. E. Grinshteyn and D. Hemenway, "Violent Death Rates: The US Compared with Other High-Income OECD Countries," *American Journal of Medicine* 129, no. 3 (2016): 266–273.
5. Centers for Disease Control and Prevention, WISQARS. A percentage change was developed using 2007–2016 age-adjusted rates for all ages.
6. A. Anglemyer, T. Horvath, and G. Rutherford, "The Accessibility of Firearms and Risk for Suicide and Homicide Victimization Among Household Members: A Systematic Review and Meta-analysis," *Annals of Internal Medicine* 160 (2014): 101–110.
7. M. Miller, D. Azrael, and D. Hemenway, "The Epidemiology of Case Fatality Rates for Suicide in the Northeast," *Annals of Emergency Medicine* 43, no. 6 (2004): 723–730; S. B. Vyrostek, J. L. Annest, and G. W. Ryan,

"Surveillance for Fatal and Nonfatal Injuries: United States, 2001," *Morbidity and Mortality Weekly Report Surveillance Summaries* 53, no. 7 (2004): 1–57.

8. Vyrostek, Annest, and Ryan, "Surveillance"; M. Miller, D. Azrael, and C. Barber, "Suicide Mortality in the United States: The Importance of Attending to Method in Understanding Population-Level Disparities in the Burden of Suicide," *Annual Review of Public Health* 33 (2012): 393–408.
9. E. A. Deisenhammer, C. M. Ing, R. Strauss, G. Kemmler, H. Hinterhuber, and E. M. Weiss, "The Duration of the Suicidal Process: How Much Time Is Left for Intervention Between Consideration and Accomplishment of a Suicide Attempt?" *Journal of Clinical Psychology* 70, no. 1 (2007): 19–24; T. R. Simon, A. C. Swann, K. E. Powell, L. B. Potter, M. Kresnow, and P. W. O'Carroll, "Characteristics of Impulsive Suicide Attempts and Attempters," *Suicide and Life-Threatening Behavior* 32, suppl. (2001): 49–59.
10. Centers for Disease Control and Prevention, WISQARS. Percentage developed using five years of most recent available data (2012–2016) and age-adjusted rates.
11. Ibid. Yearly average developed using five years of most recent available data: 2012–2016. Children defined as ages zero to nineteen; percentage change was developed using 2007–2016 crude rates for children and teens (ages zero to nineteen).
12. Centers for Disease Control and Prevention, Adolescent and School Health, Youth Risk Behavior Surveillance System (YRBSS), 2017 data, https://www.cdc.gov/healthyyouth/data/yrbs/index.htm.
13. D. Azrael, J. Cohen, C. Salhi, and M. Miller, "Firearm Storage in Gun-Owning Households with Children: Results of a 2015 National Survey," *Journal of Urban Health* 95, no. 3 (2018): 295–304. The study defined children as younger than eighteen.
14. R. M. Johnson, C. Barber, D. Azrael, D. E. Clark, and D. Hemenway, "Who Are the Owners of Firearms Used in Adolescent Suicides?" *Suicide and Life-Threatening Behavior* 40, no. 6 (2010): 609–611. The study defined children as younger than eighteen.
15. J. W. Swanson, M. Norko, H. Lin, K. Alanis-Hirsch, L. Frisman, M. Baranoski, M. Easter, A. G. Robertson, M. Swartz, and R. J. Bonnie, "Implementation and Effectiveness of Connecticut's Risk-Based Gun Removal Law: Does It Prevent Suicides?" *Law and Contemporary Problems* 80 (2017): 179–208.
16. D. C. Grossman, B. A. Mueller, C. Riedy, et al., "Gun Storage Practices and Risk of Youth Suicide and Unintentional Injuries," *JAMA* 293, no. 6 (2005): 707–714. The study found that households that locked both firearms and ammunition had an 85 percent lower risk of unintentional firearm deaths than those that locked neither.